"We wrote several patent applications using Patent Wizard for our Harmony remote control. *The Patent Writer* book follows the same winning strategy as Patent Wizard and we highly recommend it to any inventor."

—Bryan McLeod, Glen Harris and Justin Henry,
Inventors of the Harmony brand remote control

"*The Patent Writer* is the patent drafting book that obsoletes all others . . . "

—Mike Stephanos
Inventor of Duraflameís orange oil-based lighter fluid

"*The Patent Writer* is a robust how-to guide that walks you through every step to create an actual patent application. Every innovator and every Small Business Development Center in the nation should use this book!"

—Mary Wollesen
State of California Small Business Programs Director

"*The Patent Writer* clearly defines the value of combining an educated inventor with a patent attorney or agent for the winning combination for writing a successful patent application."

—Deb Hess
Executive Director, Minnesota Inventors Congress

The PATENT WRITER

How to Write Successful
Patent Applications

Bob DeMatteis
Andy Gibbs
Michael Neustel

SQUAREONE
PUBLISHERS

Every reasonable effort has been made to provide reliable information and data, but the authors, editor, and publisher do not assume responsibility for the validity of any or all materials contained herein or the consequence of their use. The content of this book is not to be considered as legal advice. If you require legal assistance, seek the advice of an attorney.

Cover Designer: Phaedra Mastrocola and Jeannie Tudor
Editor: Elaine Weiser
Typesetter: Theresa Wiscovitch

Square One Publishers
115 Herricks Road
Garden City Park, NY 11040
(516) 535-2010 · (877) 900-BOOK
www.squareonepublishers.com

Library of Congress Cataloging-in-Publication Data

DeMatteis, Bob.
　　The patent writer : how to write successful patent applications / Bob DeMatteis, Andy Gibbs, Michael Neustel.
　　　　　p. cm.
　　Includes index.
　　ISBN 0-7570-0176-9 (pbk.)
　1. Patent laws and legislation—United States—Popular works. 2. Patent licenses—United States. I. Gibbs, Andy. II. Neustel, Michael. III. Title.

KF3120.Z9D46 2006
346.7304'86—dc22

2006003651

Printed in the United States of America

10　9　8　7　6　5　4　3　2　1

Contents

Preface

Behind every great invention, the one that earns a bundle of cash for the inventor, there's a solid, smartly written patent. This statement is not simply a "motto of the month," it's our conclusion after having written scores of patents on great inventions that have earned millions for us, as well as for other inventors we've worked with, over the combined fifty years we've been in the invention and patent world.

Today, more than 7,000 patents are issued every week in the United States, and nearly three times that internationally, but only a very small fraction of them will ever earn that "bundle of cash."

Patents in business are taken seriously. There is simply too much money being placed in research and development and business investment to take patents lightly. In fact, the Licensing Executives Society, the international association of experts in patent and intellectual property valuation and licensing, says that intellectual assets, including patents, trademarks, and copyrights represent 87 percent of the Standard and Poor market capitalization. This means that intellectual property contributes more than six times to a company's stock value than all of the facilities, machinery, real estate, and inventory combined.

Poorly written patents are not only a waste of time and money, they will rarely become valuable in commerce.

On the other hand, a *properly* written patent will describe the invention using the best legal and technical terminology, achieve broad claim coverage, and close the holes that other inventors and engineers may exploit in order to earn their own patent, possibly one that's stronger than yours.

What separates a poorly written patent from a strong, defensible patent that solidly describes an invention? You're about to discover the patent writing success secrets that other patent-writing books simply gloss over—if they mention them in the first place.

JEROME H. LEMELSON

Jerome H. Lemelson was one of America's most prolific inventors. He said, "*I am always looking for problems to solve. I cannot look at a new technology without asking: How can it be improved?*" Lemelson was an optimistic and vibrant man, and his energy for ideas and discoveries continued well into his 70s. Among his scores of patents were inventions in the fields of medical instrumentation, cancer detection and treatment, diamond coating technologies, consumer electronics, and television. A few of his most notable ones are the drive mechanism of the audio cassette player, key components of the VCR, and LASER-guided industrial robots. Lemelson also believed that teaching invention, innovation, and entrepreneurship to young people is the most effective way to bring forth the next generation of innovators—generating new products, creating employment opportunities, and strengthening the national economy. With this vision in mind, his Lemelson Foundation helped create the NCIIA, an organization that supports and encourages invention, innovation, and entrepreneurship at more than 200 private and public colleges and universities throughout the United States by offering grants.

A large part of *The Patent Writer* can be attributed to two great innovators—the great novelist Oscar Wilde, and America's noted genius Albert Einstein. Wilde said "Nothing that is worth knowing can be taught." Years later Einstein echoed this notion when he said "Example isn't another way to teach, it's the *only* way to teach."

Our challenge was not to add our book to the scores of patent-writing manuals that tell you how to write a patent. Since we've already established that it's worth knowing how to write a strong patent, Wilde says that we can't teach the best patent-writing techniques. Luckily, Einstein tells us that we *can* teach you powerful patent writing, as long as we teach by example.

Since we have learned the proper patent process by writing them, then parlaying those patents into hundreds of millions in product sales, our challenge in writing *The Patent Writer* was to develop a logical, easy-to-follow format that not only taught the process of writing a patent, but one that actually applied our teachings in a real world example that you could follow from "idea to patent."

 That's why we invented the "Illuminated Hammer" expressly as a follow-along example to help illustrate the key points, important processes, and expert techniques you'll use to write strong patents. You actually follow along as we take our Illuminated Hammer from concept through patent application. This exclusive feature makes *The Patent Writer* one of the best patent-drafting references written, yet the patent-drafting process presented in this book makes it one of the easiest to apply.

Your goal is clear—you don't just want a patent, you want a successful patent that is strong enough to overcome patent office objections during prosecution, will survive infringement or invalidity attacks, and will generate enormous wealth.

Our goal is also clear—we must help you learn by experiencing the process of writing a strong and powerful patent, and through that process, spark your passion to understand, learn, and succeed!

Thanks in part to Wilde and Einstein, *The Patent Writer* is one of the shortest paths to your goal of creating that successful, valuable patent.

Introduction

Get a patent—then make MILLIONS OF DOLLARS!

It's a grand idea, and many people do end up turning their inventions into riches beyond their dreams. However, not many of those successful inventors got rich by mistake. They followed a shrewd strategy, learned the rules needed to play in the patent world, and religiously followed a proven process of writing strong, defensible patents.

The Patent Writer is about to take you step-by-step through the patent-writing process so you can start writing valuable patent applications immediately. Not just any patent-writing process, though. *The Patent Writer* leverages the authors' decades of expertise in patent law, inventing, and corporate product development. Their industry experience spans the packaging, automotive, medical, sporting goods, software, computer hardware, telecommunications, business methods, machine equipment, and many more industry segments.

So, regardless of what industry your invention is destined for, you can be confident that *The Patent Writer* will intelligently guide you by teaching a logical, easy-to-follow process of writing your own patent that's clear, focused, and strong.

If your patent is going to survive in the real world, it must be "strong." That means it must be strategically thought out in advance, then well written using the solid principles we teach throughout *The Patent Writer.*

The Patent Writer is the most up-to-date power tool currently available for those who want to learn the art of writing a solid, valuable patent. *The Patent Writer* feeds you no hype, and it doesn't sugarcoat the tough parts of writing a patent. If followed, it will enormously increase your probability of joining the other top inventors who own the patents that have earned the big money.

1

In *The Patent Writer,* you will learn:

- How to effectively write valuable patent applications.

- How to save potentially thousands of dollars writing your own patent application.

- How to write a "provisional" patent application.

- When to use patents, trademarks, copyrights, trade secrets, or other intellectual property protection for your invention.

- The different types of patents and which type you should use.

- Why the careful selection of words to describe your invention can mean the difference between a winner and loser—we'll show you how much power a single word can have.

- Why budgets are important.

- How and when to invest in the expertise of a patent attorney. Yes—there is a reason that patent attorneys spend upwards of a decade learning their craft.

WHY DO YOU EVEN WANT A PATENT?

But even before you run to your computer to start tapping out your patent, you're going to need to see the "big picture" first. We're going to take you back a few steps and make sure that you have a realistic understanding of your financial, business, and personal objectives, since they will shape your patent-writing strategy.

You probably want to make money from your patent, right? But how are you going to get someone to pay you for your invention?

Some inventors decide to go into business to make and sell their invention themselves. Other inventors have no desire to make their own products, but plan on licensing their invention to a large company that will ultimately pay them a royalty.

It stands to reason that only the inventor who wants to make and sell his or her invention needs to learn about marketing, sales, manufacturing, and finance. In reality, the inventor that plans on licensing his or her invention must become the most knowledgeable about these "big company processes."

In order to make sure your invention is successful, you must make sure that the big corporations will pay you for it. Each manager will look at your invention from completely different perspectives, and try to find reasons NOT to license it. For that reason, we'll help you to understand what motivates the finance manager, manufacturing director, marketing executive, and senior engineer, and how to make sure you write your patent to address their issues.

Your invention, and more to the point, your well-written patent, should speak to every one of these corporate managers. Your patent must answer their questions even before they are asked—and by the time you finish applying the *The Patent Writer* formula, it will.

FOLLOW OUR LEAD

The Patent Writer not only tells you the patent-writing process you'll need to follow to succeed, we immediately apply our suggestions to an actual patent application, making it easier than ever to understand how to write a winning patent. Throughout this book, you'll follow how we write our own patent application for our "Illuminated Hammer" invention.

At each critical step of the patent-writing process, whenever you see the small square hammer icon, you will see how we applied *The Patent Writer* method to our Illuminated Hammer invention. Follow the same process to writing the patent for your own invention and you are on the road to success!

In fact, our Illuminated Hammer sample will show you the entire process of taking the invention from the idea stage right through to a final patent application, including our different approaches to the invention (the embodiments), and the drafting of our patent claims to obtain maximum protection. We'll even show you the different approaches that could be used to manufacture our hammer—and you'll be amazed at how manufacturability issues actually reshaped how we wrote our important patent claims.

By following the Illuminated Hammer throughout *The Patent Writer,* you'll learn first-hand the logical process you will follow to write your own patent.

PATENT-DRAFTING TECHNIQUES

Patent drafting is a wonderful art form that you will find interesting and challenging. Inventors who draft their own patent applications are usually better educated and appreciative of the patent process. By following the guidelines in this book, you will discover the art of patent drafting and how to apply it to your respective inventions in a responsible manner.

From this point forward we will be discussing patent *drafting.* Parts of a patent are *written,* but patents also contain references to other similar patents discovered during a patent search, as well as drawings. Together, we refer to the whole of creating a patent application as patent drafting.

This book cannot show you all of the possible methods of drafting patent applications. However, in view of the much-anticipated decision in *Phillips v. AWH Corporation* (decided July 12, 2005), we can explain current laws describing how patent claims are now interpreted so you may write a successful patent application. Unfortunately, other patent-drafting books printed before the *Phillips* case utilize outdated writing techniques —and are possibly even dangerous to use. You will also want to visit www.patentwriter.com often to ensure that you are educated as to the most current patent laws and cases.

Based upon current law and practices, we're going to show you the two main approaches to patent drafting:
1. Ordinary Meaning Technique; and
2. Lexicography Technique.

You do not have to limit yourself to only one technique for drafting your patent application—you can easily choose which technique works best for each portion of your application. After reading this book, you will be able to determine which technique or combination thereof is best.

Here's a quick preview of these techniques—you'll see why it's important to understand the key differences in your writing approach.

Ordinary Meaning Technique

This technique is relatively basic and simple as suggested by its title. The patent drafter utilizes words that have established meanings to "one skilled in the art" of his or her invention. In other words, no conflicting meanings for words are used in the patent application that would give a word a meaning different than the "ordinary meaning." In addition to your written patent, dictionaries, encyclopedias, treatises, and other types of extrinsic evidence may be used by a court to help determine the ordinary meaning of the words used in your application. Throughout the book, we'll provide you with examples, making it easy for you to follow.

The potential upside to this method is that it's simple, easy, and reduces the chances of including an unnecessary limitation within your patent application. The potential downside to this method is that you are stuck with the ordinary meanings of terms as understood by a hypothetical person of ordinary skill in the art of your invention.

Lexicography Technique

Lexicography is the art of making up words or applying a certain definition to existing words. The definition you provide to words with this technique may narrow or broaden their ordinary meaning.

As stated, you are determining the exact definition of words in your patent application through this technique, which requires a solid understanding of the invention and the industry. When you specifically define the meaning of a term, you are trumping the established ordinary meaning and creating your own meaning for that term. Remember, you can use this technique for one or more of the terms used in your application while allowing the remaining words to retain their "ordinary meaning."

The potential upside to this method is that you are in control of the meanings applied to the words in your patent application. The potential downside to this method is that you may inadvertently include unnecessary narrowing language in your definitions.

Make Your Own Decision

Remember, you do not have to select just one technique for drafting your patent application. Choose the technique that you feel the most comfortable with. If you are very strong in grammar and understand the industry of your invention very well, then the lexicography technique may be attractive to you. If you are not that strong in grammar or do not fully understand the industry of your invention, then the ordinary meaning technique may be attractive to you. As always, you can mix these two writing techniques as you feel comfortable doing.

KNOW BEFORE YOU GO

We've given you the quick tour of *The Patent Writer* roadmap to writing a successful patent. Now it's time to put the process into action.

As every successful inventor knows, being prepared is the first order of business, so pull out your inventor's journal and pencil, get your computer fired up, and then get ready to begin *The Patent Writer* method! Remember, anyone can "invent." Now enter *The Patent Writer* world, and prepare to *successfully* invent—you're about to write your first winning patent.

STAY INFORMED—VISIT PATENTWRITER.COM

In addition to reading *The Patent Writer,* it is important to stay up-to-date on the latest proposed legislation, laws, and cases. Visit www.patentwriter.com often, where you will find valuable tools and updates from the authors.

Read Before Writing

You don't have to reinvent the patent-drafting process. Read The
Patent Writer, read patents, and read about technology in your field.
You will then be prepared to write a successful patent application.

This book is about making money. It is about writing patent applications intended to become economically valuable, to earn money for you and your company.

The best way to do this is to write a patent application that broadly protects your innovation, and then turn it over to your attorney to file in a timely manner. Knowledge is power. Don't go on a wild goose chase trying to patent it yourself. We'll show you a proven approach that will ensure high quality patents and is best poised for profitable results, all the while preserving cash flow.

IT'S ABOUT MONEY

The U.S. Patent system was created by an act of Congress in 1793 in order to encourage commerce in the newly formed country. The objective was to generate income and prosperity. The reward for the inventor's ingenuity was a legal monopoly and the chance to earn substantial income.

Regardless of whether you are an ongoing business, an entrepreneur about to launch a new start-up, or an independent inventor or product

**EDUCATION BEFORE
ACTION**

Many inventors make the mistake of initiating the patent process before becoming fully educated. You can literally save thousands of dollars and hours by simply taking the time to read this book!

developer who wants to license your innovations, the guiding principles of why you are writing a patent application are essentially the same.

Since successful inventors consider their patents to be valuable business tools, they decide how they are going to earn money from their patent, and build their budget to make sure that they have enough money to obtain the patent.

There are two methods of generating revenue from your patent: license your patent to another company that will pay you a royalty for the use of your patent, or manufacture and sell your patented product yourself.

Patent to License

The "formula" for a strong patent might change slightly depending on the inventor's objectives. On occasion, a patent that an inventor intends to license to industry may be written a little differently from a patent intended to protect a product that inventors will manufacture and sell themselves.

Writing a patent with the primary objective of licensing it to a manufacturer, distributor, or other licensee requires attention to a few details of particular importance to licensees (your "customer").

A licensee would define a "strong" patent as one that includes these attributes:

- A licensable patent that protects a discrete invention. This patent does not necessarily disclose more than you want the licensee to know about your future improvements, and may prevent broad interpretation that would allow the licensee to expand the product line outside of your intended scope of protection.

- A patent application that is timed to preserve international patenting rights for the licensee—especially if it is a multinational company. Even though you may have no intention of selling your invention internationally, large corporations may. Preserving *their* ability to file international patents can create a higher licensing value for your patent.

- A patent that discloses multiple inventions, laying the groundwork for future continuation-in-part (CIPs) or divisional patent applications, may also protect your ability to improve the invention based upon the licensee's market thrust. It may also provide you with a means to develop and patent the invention in other fields that are not related to your licensee.

- The patent must protect all aspects of the invention pertinent to the licensee's field. If it doesn't, it may be vulnerable to other patents and products using similar methods and systems. At times your patent protection may include both the "best way" as well as the "second-best way" to make or use your invention, thus preventing inferior (and perhaps lower cost) copycats from entering the market.

Generally speaking, patents written for the specific purpose of licensing must be just as strong as those that a company would write to protect its own technology in its business. However, if a company is already established with an existing product in the marketplace, it may wish to develop minor improvements specific to its product line, resulting in narrower patent protection. Typically, an inventor would not file such a narrow patent to protect these small product improvements, since the licensing revenue potential would be very small.

Patent to Make and Sell

Patents developed by shrewd inventors/entrepreneurs may differ from patents written for licensing. One way they may differ is that such patents may disclose a broad range of inventions, as compared to the more specific scope of a patent to license.

By employing patent strategy and tactics, this kind of inventor may find sensible ways to conserve cash now, while preserving the opportunity to later file multiple patents on many new inventions or improvements.

Important features of applications for patents the inventor will make and sell may include:

- Writing a provisional patent application to preserve the filing date (known as the priority date).

- Writing an application that has an early filing date that reserves the right to file in foreign countries.

- Writing an application for a Patent Cooperation Treaty (PCT) filing (see page 161–162).

- Writing a patent application that leverages the PCT filing during the

9

national phase patent examination—or the phase at which a PCT application is filed for a United States patent (see page 188).

All of this may sound complicated, but these are important considerations that you will become more comfortable with by the time you have finished *The Patent Writer*. These are but a few patent strategies and tactics that must be considered during the writing of an application for a strong patent. Regardless of your approach and your strategy, it's important to learn all you can first, and then consult with your patent attorney to determine the right strategy for your needs.

Patent Budget

If you don't plan on making any money from your patent, you are well within your rights to ask your patent and legal professionals to give you their time and expertise work for free.

PROSECUTION

Prosecution is the negotiation between the patent examiner and the applicant on the allowance of the claims during the patent pending period.

Before you laugh at that statement, you should know that the authors, who work with literally thousands of inventors and small businesses every year, see many inventors who time and again, expect patent attorneys to jump on their idea for a piece of the action, or for the patent attorney to wait for payment for legal services until after they make their "millions."

The commercial reality is that the more valuable your patent becomes, the more you will be investing in writing, protecting, and enforcing it. But the big question is "how do I know how much to invest before I make anything from it"? It's like asking how many bingo cards you should buy, not knowing whether you will win.

The simple answer is "we can't tell you." What we can tell you is that you must get a handle on what the typical costs are to file, prosecute, and have your patent issued.

For discussion purposes here, let's assume that the average costs to have a patent attorney write the final patent application (writing or rewriting your draft to make it stronger, and prosecuting the application through until it issues) are as follows: simple mechanical patent: $5,000; complex plant, chemical, or software patent: up to $15,000.

EXISTING COMPANY

If you're an engineer or intellectual property manager with an existing company, you may wish to introduce new product lines or improvements on existing products in order to grow and expand the company.

Your obligations to the company's shareholders include protecting your company's assets and profiting from proprietary product line. Smartly drafted patents provide a sound means of helping you meet these obligations. However, if patent applications are filed with a shotgun approach without validation of marketability of the products they protect, it begs the question, "Is this a responsible approach?" We believe it is not.

We don't intend to get into the specifics of your company's patent strategy, but if you're a competent engineer or intellectual property manager, we're confident that pursuing *The Patent Writer* approach to patent drafting is a smart decision. In a corporate world filled with profit-making obligations to the company and its shareholders, this approach only makes sense.

You may be surprised to find that patentable subject matter is found in virtually every department in your business—manufacturing, engineering, marketing, maintenance, and yes, even in your accounting department. You should be aware of the role of patents throughout your organization, which we'll explain in detail later on.

Are you preserving your company's innovative marketing methods, or are you letting competitors copy them? Are you patenting vulnerable trade secrets, or are you risking having competitors independently discover them, only to use them against you in the near future? Are you using some accounting or business methodology that has been proprietary, yet unprotected? And what will you do when some other entity files a patent application on it?

A company's interests should be protected with patents in every department so that it may continue to prosper and generate ongoing revenue. By doing so, a company is illustrating the highest degree of shareholder responsibility and securing its future as well. From this perspective, *The Patent Writer* is an invaluable aid to all individuals in your company who may be enlisted to write a patent application on their inventions.

POORLY WRITTEN PATENTS = NO VALUE

It is estimated that approximately 95 percent of all patents fail to return more money to the inventor than its cost to obtain the patent in the first place. Commercial value of these patents is usually minimal, based partly on the fact that they have poorly written (weak) claims.

ENTREPRENEURIAL START-UP

If you plan on starting up a new business based upon an innovative product or technology, the best initial market protection you can obtain is through patents. It makes no sense to launch a new company only to have larger competitors muscle you out of your market once you've proved the profit potential.

For an entrepreneurial start-up, the best way to secure strong patent protection is through thoughtful consideration of the patentable subject matter, and working with your attorney to determine a sensible patent strategy that fits your budget. After all, your finances are best spent on marketing and ramping up the new innovations—not filing a bunch of patent applications that won't directly contribute to your company's value.

INDEPENDENT INVENTOR OR PRODUCT DEVELOPER

If you are an independent inventor or product developer and plan to license your patents, you'll have to have broad enough protection to protect your licensee's interests. Nothing may be more important to your success than this fact.

For example, you may not have the expertise to produce high quality prototypes or working models. You may not have the financing or expertise to launch a small business or to conduct product testing to verify the inventive aspects of your invention. But if you have strong patent protection and qualified marketability, you can bet that licensees will be eager to license your patent and to finish up development.

Above all, we sincerely hope that if you are an independent inventor, you do not go chasing after a false dream thinking that a single, narrowly written patent is all you need to become the next millionaire. It rarely happens, if ever. You would be better off investing the money in the lottery.

On the other hand, follow the tenets in *The Patent Writer* and you'll be able to write successful patent applications that are more valuable to both you and future licensees. It is perhaps more important that you to learn to write a successful patent application to protect *your* interests first. Why?

First, a high percentage of new inventions over the past two hundred years were developed and patented by independent inventors—Eli Whitney's cotton gin, Edison's light bulb, Yamamoto's cordless telephone, and a whole lot more.

There are two more important reasons you should follow *The Patent Writer*'s format. First, it'll save you a lot of money during the patent-drafting and development process. Second, many brilliant inventors unfamiliar with patent laws and modern patent-drafting methodologies have tried to write and file patents themselves, resulting in patents with such narrow scope that they end up being commercially worthless. After realizing that all their efforts were for naught, many never invent or patent again.

INVENTION PROMOTION COMPANIES

Invention promotion companies are a travesty to independent inventors. Not only can they prematurely expose an innovation causing you to lose your rights, the quality of their patent searches and patent applications often lead to patents that are not enforceable, and of little commercial value.

The Patent Writer is a powerful tool that helps inventors understand the important approaches to writing every element of a patent application to best ensure that they receive the strongest, most defensible patent possible. *The Patent Writer* helps inventors broaden the scope of patent protection, and increase their odds of being successful.

SELF-DRAFTING ADVANTAGES

There are several advantages to drafting your own patent applications.

Save Time

You can file your patent application with the United States Patent and Trademark Office (USPTO or just PTO) sooner if you write it yourself before handing it off to your patent attorney. The conventional route is for the inventor to provide an initial disclosure to a patent attorney and then wait for the patent attorney to prepare the application. If you want to be patent pending quickly, we'll show you how to best write your own application before giving it to your attorney to finalize and file.

Save Money

Time is money, and patent attorneys typically charge $5,000 to $7,500 for the time it takes to prepare a utility patent application for filing. Writing your own patent application will save countless hours in teaching your attorney about your invention, and in the process can save you thousands of dollars in attorney fees.

Improve Patent Quality

You will find that by writing your own application and putting your invention to paper, you will continually refine, broaden, and improve your invention—and the quality of your application.

Identify Design-Arounds

When writing your own patent application, variations on your invention will emerge. These variations are potential *design-arounds* that could be exploited by competitors if you do not adequately protect them in your patent application.

SELF-DRAFT?

Self-drafting a patent application is an experience few inventors ever regret—however, filing and prosecuting a self-drafted application is another story. Each invention must be evaluated with respect to the risks and benefits of self-drafting without utilizing the services of a qualified patent attorney.

13

**ALTERNATIVE
EMBODIMENTS**

It is important to start
thinking about *all* possible
ways of making and using
your invention—including
alternative embodiments
that may not have
significant marketability
in your opinion.

You will want to fully disclose all possible variations of your invention in the patent application so they may be covered with broadly written, strong claims. This serves to turn your patent into a defensive document as well.

Improve Claim Quality

The claims of a patent application are one of the most important sections. By writing a patent application yourself, you will develop a broader claim strategy for your patent attorney to pursue.

WHO WRITES . . . WHO FILES?

There are three approaches to writing a patent application, assuming that you're doing the writing. Of course you can bypass the entire process and have your attorney do it all for you. However, that approach tends to produce inferior or more costly patents for the reasons previously discussed.

The Best Strategy

The preferred and safest strategy for writing your own patent application is to educate yourself about the patent process; write a provisional patent application; and have your patent attorney review, edit, and file it with the USPTO.

Within one year after the provisional application has been filed, have your patent attorney prepare and file the "permanent" patent application (sometimes called the "regular" or "non-provisional" patent application) that contains the legal claims. It is then the job of your patent attorney to prosecute this application with the patent office to ensure that you receive the broadest possible patent protection.

This is the best approach for businesses, inventors, and engineers who are not experienced patent drafters, but who want to make sure their applications are accurately written, and cover all of the important inventive matter.

The Next-Best Strategy

The second-best strategy is for you to write and file the provisional application with the USPTO. After you have filed the provisional application, your patent attorney will prepare the permanent application within one year. If you have prepared a solid provisional application, your patent attorney may use your entire provisional application, add the legal claims, file the application, and prosecute it.

Those who are experts in their field, know the subject matter well, and are knowledgeable on patent drafting, commonly use this approach. It becomes an easy task to take an existing patent application and modify it to accompany the new inventive matter in a new patent application.

Should you become adept at patent drafting and use this method, send a copy of the provisional application and the filing date as soon as possible to your patent attorney so he or she may put it on his or her docket. Usually, about nine months later, you'll get a call discussing the claims for your permanent application.

The Do-It-Yourself Strategy (*Pro Se*)

It is difficult for average inventors to get the strongest and most defensible patent by filing and prosecuting a permanent patent application themselves.

We strongly recommend against taking the "do-it-yourself" approach to patenting your invention. Patents are legal documents, and the process of writing and prosecuting patent claims requires the knowledge and training of a professional patent attorney. We know that many inventors have filed their own applications and have received patents. However, we don't know very many inventors who have successfully defended their self-filed applications in a patent infringement suit.

WHAT YOU SHOULD KNOW BEFORE YOU START

Whether you are an employee within an existing company, an independent inventor/developer forming a start-up, or you plan to license or partner your innovation, there are a few essentials you'll want to know about before you begin.

Don't take the "ready-fire-aim" approach to patenting used by many first-time inventors. That is, rush to file a patent application, then try to learn about the process, and finally hope that there will be a market for the innovation. That's backward . . . and invariably a costly, time-consuming mistake.

DO YOU PERFORM YOUR OWN SURGERY?

The most successful inventors and small business entrepreneurs focus on their invention, and let their patent attorneys focus on making their patent documents and claims as strong as they can be. Draft the patent yourself, then use your patent attorney to clean it up, and build a strong claims section.

In order to construct a strong patent position, you must first become adept at the terminology in your field and understand your patenting options. Let's highlight a few important points.

First-to-Invent Laws

The United States presently determines priority among competing inventors by reference based on who was the *first to invent*. Except for the Philippines, all other countries determine priority by who was the *first to file* the patent protection.

In other words, the inventor who can prove his or her invention was reduced to practice first has priority over another inventor who made and proved the same discovery later on. There is a caveat however, in that the inventor must exercise diligence in reducing an invention to practice, and then subsequently patenting it. The best way to establish your first-to-invent rights is by keeping your records in a journal.

As of the publication of this book, the United States had proposed legislation that would change the U.S. Patent System to a *first-inventor-to-file* system. Make sure to visit www.patentwriter.com frequently for legislation updates.

Use a Journal

Proper record keeping is an important first step to show proof of the date of original conception (a.k.a. the date of invention), and the date in which your invention was reduced to practice. Reduction to practice typically occurs by properly illustrating in your journal how your invention works. It may also occur by building a working prototype or by the filing of a patent application. Obviously, specifications and notes written in a journal will establish a priority date over the more time-consuming task of filing a patent application.

You should also know that mailing a sealed letter with invention documents to yourself is not an accepted method of proving your invention's conception date. Using the invention disclosure and then recording your day-to-day activities in a *scientific journal* is the preferred way to maintain records on your invention.

More importantly, your journal records will reveal the inventive matter on which you'll be writing a patent application. You'll record what

---

THE VALUE OF JOURNALS

A detailed journal not only proves your "invention date," it also helps to identify potential co-inventors that need to be named in a patent application. A journal can also be a good resource for your patent attorney in preparing a patent application.

INVENTOR'S JOURNAL

Record your day-to-day activities in a scientific journal in which the pages are pre-numbered and permanently bound so that material cannot be added or removed later on. This makes it a legally defensible diary proving the evolution of the invention-development process.

materials you used, prototype testing records showing what worked and what did not, as well as your drawings and sketches.

Your journal also proves your continued efforts to reduce your invention to practice, a Patent Office requirement.

Marketability

Marketability of an invention is obviously important. If an invention is not marketable, there is no need to file a patent application. Have you fully researched and qualified a solid market opportunity? If not, you may be pursuing a false dream believing that your patent is going to be valuable.

A marketability evaluation is what all inventors should complete prior to attempting to market their invention. A marketability evaluation basically considers whether the invention is marketable within the current and future market. There are a number of organizations that provide legitimate evaluation services for hire, or consider completing your own evaluation using the online evaluation program at www.patentcafe.com.

ILLUMINATED HAMMER

Here is an example of two of the initial drawings we recorded in our journal on our hammer invention. Inventions invariably go through a metamorphosis. Expect yours to do the same. In Chapter 8, you'll see the end result and what we filed our patent application on.

Manufacturability

Just like marketability, manufacturability plays an essential role in developing a successful invention. Take the computer industry, for example. Even though more and more features (improvements and inventions) have been added to the personal computer over the years, retail prices continue to fall, not increase. Should an improvement to a computer cause prices to increase instead, it may not be accepted by consumers.

It's a well-known economic fact that price elasticity (the extra price consumers will pay to replace an existing product they buy) is generally no more than about 2 percent for most consumer products. If the cost of your innovation is going to increase the sell price by 10 percent, how will it be received in the marketplace? If your innovation is going to increase costs by more than 2 percent, you may need to

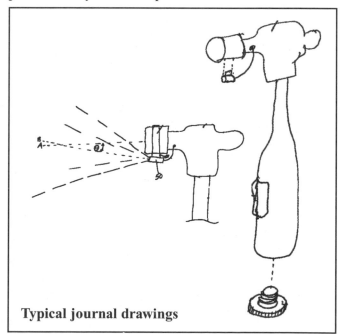

Typical journal drawings

innovate a new production process as well, in order to be competitive. Otherwise, the tiny market your innovation may capture (probably less than 1 percent) if any at all, may not be worth the cost of patenting.

Patentability

One of the first steps of the patenting process is to conduct a patent search to determine the likelihood that your invention is patentable. You can perform preliminary patentability searches on the Internet using the multi-country patent database at www.patentcafe.com, the software program at www.patenthunter.com, or connect to the United States and European patent office databases.

After performing a preliminary patentability research, you will want to have a patent attorney provide you with a professional patentability opinion based upon an independent patentability search.

Patent Drafting and Development Strategy

Patent drafting and product development go hand-in-hand. File your patent application too early in the process and you could lose priority for patentable improvements found only after you are in the thick of product development. You may ensure that your product development is on track with your patenting, manufacturing, marketing, and licensing activities (as the case may be) by reading the invention development guide, *From Patent to Profit* and using its "Strategic Guide Chart" to track your progress.

Coordinate your patent drafting along with prototyping and product development to ensure that you are maximizing the patentable subject matter and creating a strong patent application.

THE PATENT WRITER PLAN

The Patent Writer plan takes you through all the steps necessary to draft a successful patent application.

Follow this simple plan as you read through the chapters. It sets forth a proven patent-drafting method used by many successful inventors and patent attorneys. You may refer to this plan as a roadmap. In order, here's your nine-point checklist:

1. **Learn.** Learn about patenting and patent laws.

2. **Record Invention.** Use a journal to establish your first-to-invent rights, and maintain records, sketches, and invention and patent-related subject matter.

3. **Patentability.** Do a patent search and qualify patentability.

4. **Marketability.** Qualify marketability and manufacturability of the invention.

5. **Identify Patentable Subject Matter.** Complete the *Patentable Subject Matter Checklist* on page 86.

6. **Claim Statement.** Prepare a claim statement.

7. **Describe Invention.** Follow the instructions in the book and use our patent-drafting terminology along with words commonly used in the field of your invention.

8. **Patent Drawings.** Prepare (or have prepared) drawings using the formats illustrated in Chapter 9.

9. **Strategy.** Follow through and file the patent application based on the patent strategy that's right for you and your invention.

These are the important points—the important steps you'll be taking to turn your inventions into successful, valuable patents. Follow them as you read the chapters in *The Patent Writer* and you'll soon become an expert in patent drafting and an expert in your field.

USEFUL TOOLS

There are many useful tools for inventors during the early stages of invention development and patenting. Following is a list of commercial products developed by the authors that can assist you with documenting your invention, determining marketability and patentability, and preparing your patent application. For more information on these, as well as helpful governmental and non-commercial resources, see the Resource List on page 225.

- *Scientific Journal*—Using a journal is the best way to ensure that the concept date of your invention, and any improvements made during development, are properly documented. Using a journal is the fastest and least expensive method of reducing your invention to practice and establishing your first-to-invent rights.

THE BEST CRAFTSPEOPLE USE THE BEST TOOLS

Don't try to go it alone—the best tools to help you with your patent drafting are only a click away.

- *From Patent to Profit*—The invention development guide for small business and independent product developers is a rock-solid reference companion that teaches market research, budgeting, patent licensing methods, and much more.

- *PatentCafe's Invention and Marketability Software*—The only confidential, online evaluation software that produces a complete report highlighting strengths and weaknesses—and suggests ways to overcome the weaknesses.

- *PatentWizard*—The industry's premier software for writing a provisional patent application. With your input, the software creates an application and the necessary forms for filing it with the USPTO.

- *PatentCafe's ICO Global Patent Search*—The most advanced patent database search engine lets you search patents using natural language technology developed for the intelligence community.

- *PatentHunter*—An advanced program that allows patent attorneys, businesses, and inventors search, download, and manage United States and foreign patents.

- *Patents in Commerce*—Learn about innovation and how to get free government assistance from more than 1,000 SBDC (Small Business Development Center) offices nationwide.

Now that you have a broad understanding of how to write a powerful patent, you're ready to start on your own invention, right? Maybe patents aren't all you need to protect your invention. There are a few other forms of intellectual property protection that you must also understand. Sometimes, a patent is only a part of a more aggressive protection strategy, and the patent will need to be written as a part of your overall plan. In Chapter 2, we'll take a quick tour through other forms of intellectual property to help you choose the best options for protecting your invention.

Patent Basics

Patents are intellectual property. Before you write your application,
you must understand the various forms of patents and related forms
of intellectual property and which may be applied to your invention
to obtain the strongest protection.

Your objective is to write a successful patent application and secure
bulletproof protection. Before you begin writing, you should have
good working knowledge of the various forms of patent protection and the
various forms of related intellectual property that may apply to your inventive endeavor.

Securing strong intellectual property protection on your invention frequently overlaps into your ability to secure additional scope via trademarks, copyrights, trade secrets, and mask works. By understanding these
forms of protection you will broaden the scope and value of your intellectual property.

TYPES OF INTELLECTUAL PROPERTY

There are five forms of intellectual property (IP) you should consider in
your invention development—patents being just one of them and usually
the most important. The five key types of intellectual property protection
you should consider are:

PATENTS ARE VALUABLE

You will hear some
inventors tell you that you
do not need a patent to
license the rights to your
technology or to
manufacture your product.
Successful inventors know
that patents are required to
successfully license the legal
rights to the technology and
to keep competitors at bay.

- Patents
- Trade secrets
- Trademarks

- Copyrights
- Mask works

Patents

Patents protect the *functional* aspects of your invention (i.e. utility patents) including products, processes, systems, and so on, and at times the *ornamental* appearance (i.e. design patents). You may file for both utility patent protection and design patent protection on the same invention. Patents give you a legal monopoly for the term of the patent. With a few exceptions, patent terms are: twenty years from the filing date for utility patents, and fourteen years from the filing date for design patents. You'll read more about this later.

Trade Secrets

Trade secrets define subject matter that conceals anything used in the course of business to provide a competitive advantage. It is subject matter that is *not disclosed* in a patent application or the final granted patent document.

A trade secret is only that subject matter that you believe may be held confidential for the long term. For example, the formula to Coca-Cola® or the recipe for Kentucky Fried Chicken® are better held as trade secrets because if they were patented, they'd be publicly disclosed in the patent documents. Upon expiration of the patent's term, anyone may practice or use what has been disclosed.

If trade secrets become public knowledge, there is no way to go back and protect them with patents. Intellectual property protection is lost. If you have doubts about keeping a trade secret confidential, be prudent and patent the subject matter. Trade secrets are commonly used in the form of manufacturing processes, formulae, marketing data, and so on.

Trade secret subject matter may be used by your company or licensed along with an issued patent. When doing so, it's important to make sure that your employees and/or licensees gain access to the trade secrets on a need-to-know basis. Companies should have an employee confidentiality program in place and licensees should operate under strict confidentiality agreements. Can your trade secret subject matter be safeguarded in such a

---

TWO BETTER THAN ONE

Many inventions are covered by more than one form of intellectual property beyond patents alone. Discuss with your attorney how copyright, trademark, or trade secret protection can help protect your invention in other important ways.

manner? Use this question as a litmus test to qualify whether or not trade secret protection should be sought.

Trademarks

Trademarks identify the source of goods or services. Typical examples are the Coca-Cola® brand name or its red-wave insignia. Service marks are trademarks that describe the origin of goods sold, such as Wal-Mart's "We sell for less always.ᔆᴹ" If a product design or shape has acquired distinctiveness in the marketplace, a trademark may be acquired for such (e.g. shape of the classic Coca-Cola® bottle—U.S. Trademark Registration No. 0696147).

To register a mark, a trademark application is filed with the U.S. Patent and Trademark Office. Until you receive acceptance as a registered trademark you can use the common law notice by adding the letters "TM" next to your logo or mark. It is illegal to use the "circle R" on your product until registration has been accepted. Registered trademarks last indefinitely providing they are renewed every ten years.

Copyrights

Copyrights provide the exclusive right to copy literary and artistic expressions. Copyrights may also be used to protect software code and programs, photographs, theatrical productions, even marine hull designs, or your family movies. Copyrighting the various aspects of your innovations is common and should be considered in your product's development. For example, you may copyright the package design (such as the artistic packaging used on Celestial Seasonings® Tea boxes), any artistic copy your product uses (cartoons, photos, and so on), patterns and plans, and manuals and computer software code you've created to illustrate your invention's function and performance.

Copyrights are often used by game inventors. Patents may protect the functionality of a game, whereas copyrights may protect the appearance of the board it uses and the instructions.

Copyrights may be registered at the U.S. Copyright Office and last for the life of the author plus 70 years. Copyrighted material such as that used in package design may also be registered in the form of the trademark protection (trade dress). Your attorney will help you qualify the best approach to protecting your designs.

Mask Works

A special type of intellectual property protection is made available to the semiconductor industry.

The Semiconductor Chip Protection Act (SCPA) of 1984 established a new type of intellectual property protection for mask works that are fixed in semiconductor chips. It was promulgated as Public Law 98-620.

Mask works are not protected under copyright law. The Protection under the SCPA extends to the three-dimensional images or patterns formed on or in the layers of metallic, insulating, or semiconductor material and fixed in a semiconductor chip product (i.e. the "topography of the chip"). Although these images or patterns are purely functional features, they are nevertheless protected, provided that the particular mask work is neither dictated by a particular electronic function nor is one of only a few available design choices that will accomplish that function.

To be protected under the Act, mask works must be the original work of the author.

Summarizing Intellectual Property Types

An easy way to visualize the five basic types of intellectual property is to think of your personal computer. Utility patent protection would apply to the functional features of the electronic circuitry, keyboards, the mouse, how the computer works, and so on. Design patent protection would apply to the ornamental appearance of the computer such as the shape of the housing. Apple computer protects the design of its iPod® with design patents. Trade secret protection might apply to the manufacturing processes a manufacturer applies in order to undercut a competitor's price. Trademark protection would apply to the brand name of the computer (e.g. Apple® or iPod®). Copyright protection would apply to the software on the hard drive, the promotional material, the written manual, and any related written or visual works. Mask works protect the design of the semiconductors such as the microprocessors soldered on your computer's circuit board.

PATENT DEFINITION

In simple terms, a patent is a document granted by the United States government to an inventor giving him or her the right to *exclude* others from making, using, selling, or importing an invention in the United States. The inventor may license or sell the rights described in the claims of the patent.

There are currently over seven million United States patents issued to inventors. Without a patent, anyone can make and sell your invention without your permission and without compensation. For this reason, securing broad, strong patent protection is your primary objective. It typically serves as your first defense against competition while you establish your product in the marketplace. Over time the associated IP, such as trademarks and copyrights, will also grow into valuable assets.

PATENT BENEFITS

Protecting inventions with patents provides many benefits for businesses and inventors. The chief benefits include:

1. **Exclusive rights** that allow the patent owner to exclude others from making, using, or selling the invention. You have a legal monopoly while you ramp up sales and penetrate markets.

2. **Selling or licensing** the invention. Without patent protection, there is usually very little, if anything at all, to license. A company that is interested in purchasing or in licensing your patents will want substantial protection. Thus, this is your chicf objective and the principle purpose of this book.

3. **Increasing the value of business** to potential purchasers and lending institutions. If you plan to use your patents to expand your present product line or plan to go into business, a portfolio with broad patent protection will substantially enhance the value of the company. If you plan to have a public offering or secure venture capital, broad protection will be a crucial consideration.

4. **Increasing the marketability** of products covered by patents since consumers like unique products. Almost every consumer wants value added products that improve his or her lifestyle and improve the workplace. Protecting these valuable marketing benefits is common sense.

In summary, you may say that patents may protect the sales of your products, thus allowing you to be the exclusive profiteer. However, if not well thought out, there may be no protection at all.

RIGHT TO EXCLUDE

It is important to note that a patent does not give you the right to manufacture your invention. A patent only provides you the right to *exclude* others from making or selling an infringing product. You still need to make sure that your product is not infringing upon another patent.

MYTH: 10 PERCENT THEORY

Some inventors mistakenly believe that a patent can be "designed around" by simply changing the protected invention by 10 percent. This is not correct, as you will learn later on.

ILLUMINATED HAMMER

In Chapter 8, we write a utility patent covering at least four of the inventive attributes in our invention. In Chapter 10 we also file an inexpensive design patent on the appearance of the preferred version. This is the type of simple strategy you will learn in *The Patent Writer* and may apply to your inventions.

TYPES OF PATENTS

There are three types of patents you can pursue. Generally speaking, the broadest protection is in the form of utility patent protection, which is the central focus of *The Patent Writer.*

Utility Patents

A utility patent protects the *function* of an invention. Utility patents are granted for any new, useful, and non-obvious process, machine, manufactured article, composition of matter, software, or any new and useful improvements to an invention, product, or process.

The term of a utility patent is twenty years from the date of *filing.* Since utility patents are usually more desirable than design patents, this should be the central focus of your patenting activities. Recently enacted laws allow patent terms to be extended depending on delays on the part of the Patent Office. Your patent attorney will inform you if your patent becomes subject to patent term adjustment.

Design Patents

A design patent protects the appearance of an invention and is granted for any new, original, and ornamental design for an article of manufacture.

The term of a design patent is fourteen years from the date of *issuance.* A design patent should only be chosen if the appearance of the invention is important; otherwise utility patent protection should be sought. As with utility patents, patent term adjustment may apply to design patents.

An excellent example of the use of design patents is those used by Harley Davidson with its new V-Rod brand motorcycle. Design patents cover the motorcycle and several components. However, Harley Davidson has secured substantial utility patent protection as well.

In Chapter 10, we discuss how to properly prepare a design patent, should this be a consideration with your patent strategy.

UTILITY OR DESIGN?

Many inventors make the mistake of focusing on one type of patent protection. In fact, many inventions may be protected by both utility and design patents. Design patents have the benefit of being granted earlier and are generally cheaper to enforce.

Plant Patents

A plant patent is granted to anyone who invents or discovers and asexually reproduces any distinct and new variety of plant other than a tuber-propagated plant.

The owner of a plant patent shall have the right to exclude others from asexually reproducing the plant or selling or using the plant so reproduced for twenty years from the date of filing.

Excellent examples of plant patents include the Chandler Walnut; Flavr-Savr engineered tomato variety; the patented Chandler, Camarosa, Douglas, and Selva strawberry varieties; Madame Alfred Carriere rose; as well as many patented tulips, blueberry, and blackberry varieties. It would be highly uncommon to secure utility or design patent protection in conjunction with plant patent protection.

UTILITY PATENT APPLICATIONS

There are two main types of utility patent applications: (1) the provisional application; and (2) the permanent application. The provisional application has the same legal requirements of disclosure as the permanent application, except that claims are not required.

Whether you write a provisional or permanent application first depends on your invention development and patent strategies, as well as your patent-drafting skills.

In either case, your ability to write a patent application so your legal counsel may review and edit it is an important first step. The strongest patent applications (and ultimately the better issued patents) are usually those that are drafted first by the inventor.

Your Patent-Drafting Strategy

If you have never written a patent application, we recommend starting with a provisional application because you do not have to deal with the relatively complex claims section nor do you have to prosecute your patent application in front of the USPTO. However, we still recommend having a patent attorney review your provisional patent application prior to filing. Helping you write a provisional patent is a primary objective. This is almost always a wise first step towards securing superior patent protection, all the while conserving time and cash flow.

As you become more advanced with your patent-drafting skills, you will want to consider venturing into preparing permanent patent applications with claims. Except in rare occasions, it is strongly recommended to always have a patent attorney file and prosecute a permanent patent application.

ILLUMINATED HAMMER

At first we realized that our hammer had patentability as a device (or product). But as we developed the concept, we found additional, and perhaps more valuable, patentable subject matter. As you read the following pages, make sure you list the patentable subject matter that pertains to your invention. It doesn't cost much more to include it in your patent application and can significantly broaden your protection.

Trying to patent it yourself invariably results in poor to mediocre patents having a very narrow scope. You'll have wasted a lot of valuable time and money on securing weak patent protection that no one will want to license, partner, or fund the development.

PATENT SCOPE

The scope of protection you receive with your patent will depend upon two significant factors:

1. The amount and manner in which you disclose patentable subject matter; and

2. How you claim the patentable subject matter.

A mistake in disclosing or claiming the patentable subject matter can result in a patent with limited protection. Hence it may be worthless, easily allowing competitors to design around it. It is important to take your time in preparing an application, and do not rush into the process without first properly educating yourself—or worse yet, without properly identifying the inventive subject matter! Unfortunately, it happens all the time.

Disclosing the Patentable Subject Matter

BECOME A WORDSMITH

In Chapter 7 we'll show you how to use words and language to broadly describe and disclose a multitude of variations of an innovation by using only one or two examples.

It is important to fully disclose your invention with significant detail (i.e. structure, functionality, and operation) so that one skilled in the art will fully understand how your invention operates, attempting to keep a relatively broad scope as permitted by the prior art. This includes disclosing all possible variations of your invention, regardless if you plan to actually use the variations or the feasibility of the variations.

You should include drawings of the various features and alternative embodiments of your invention so they are clear and concise and sufficiently depict the inventive matter. Professional patent drawings are strongly recommended if you do not have the essential skills to illustrate your invention. If so, you will want to employ a draftsperson skilled in preparing patent drawings.

This reinforces what we previously mentioned with regard to writing your patent during the product development phase. This is when you will find additional, and often critically important, features or other patentable matter that should be included in your application.

Claiming the Patentable Subject Matter

With utility patent applications, it is also important to properly draft your claims without unnecessary limitations. Legally, the claims define the invention protected by the patent. From a practical standpoint, the claims determine the scope of the protection awarded by the patent office. When patent infringement occurs, it's the claims that must be infringed. Hence, the claims will ultimately play a key role in determining the patent's value and influence.

Drafting the claims of a patent application typically requires a professional patent attorney who has drafted hundreds of patent applications for other businesses and inventors.

You'll learn all about claims and their importance to your protection in Chapter 6. From the perspective of writing strong patent applications, it is important to understand what claims are and it's essential that you qualify and prepare a claim strategy before you begin writing. By doing so you'll be incorporating the most relevant patentable subject matter in the application.

When preparing your first patent applications, try to have a patent attorney assist you with the drafting process to avoid making costly mistakes. It makes no sense to file a patent application that will result in a very narrow scope or is easily designed around. It will only result in a huge waste of time and money—and can frankly have a horrible adverse effect on the product line you are about to introduce. Narrow claims effectively enable your competition to produce similar products, rather than prohibiting them.

WHAT IS PATENTABLE?

In the language of the patent statute of the United States (35 U.S.C. §101), any person who "invents or discovers any new and useful":

1. **Process;**

2. **Machine;**

3. **Manufacture;**

CLAIM STRATEGY

You should first identify the most marketable features of your invention. Then determine what type of patent protection, if any, you can receive for these individual features.

GET AN OPINION

You should have a good understanding of what is patentable after reading this book. However, it is still important to have a patentability opinion prepared by a qualified patent attorney.

ILLUMINATED HAMMER

During development we found that the initial concept of attaching a light to the head of a hammer was not feasible. As we developed the internal light emitting system, we had to make several new discoveries to ensure it was feasible. These include a method of manufacture and a composition of matter claim on the best material to manufacture the device.

4. **Composition of matter;**

5. **Method of doing business;** or

6. **Improvement** of any of the above;

may obtain a patent subject to the conditions and requirements of the law. After much litigation, patentability has also been established for genetically engineered microorganisms, computer software-based inventions, and methods of doing business, such as the infamous Amazon.com "One-Click" patent (U.S. Patent No. 5,960,411). Throughout the book, you'll come to understand these categories better, and how you can apply them to your patent-drafting strategy.

The term *process* is defined by law as a process, act, or method, and primarily includes industrial and technical processes. An extreme example of a patentable process is a method of folding a blanket (U.S. Patent No. 5,056,172). A recent explosion of patent applications have been filed and issued relating to "business method patents" since the *State Street Bank* case in 1998.

The term *manufacture* refers to articles that are made and includes all manufactured articles. Examples include shoes, chairs, computers, machinery, toys, and pens.

The term *composition of matter* relates to chemical compositions and may include mixtures of ingredients as well as new chemical compounds. This may also refer to the composition of a manufactured article when made with a certain combination of raw materials, for instance blended plastics.

In Chapter 5, we'll discuss in length how to incorporate the various forms of inventive matter in your applications and broaden the potential scope of your patent protection. What's important to know now is that you understand that patent protection does not have to be limited to just a simple "product." You can also patent how it's made, what it's made of, and how it's used. Smart inventors know how to incorporate these various forms of patent protection in their applications, which may not increase filing costs. You'll want to do this whenever you can, since your primary objective is securing broad patent protection . . . not just "getting a patent."

Novelty

To be patentable, an invention must be considered new or novel. It is considered novel providing the following conditions are met:

- The invention was not known in any part of the world at any given time before you came up with the idea.

- It was not previously described in an article and published anywhere in the world.

- It was not previously patented anywhere.

- The difference between your invention and a previous patent (or publicly known product, process, etc.) is such that it would not have been obvious to any person skilled in the art. For instance, simply changing size or color will not be acceptable.

- Your invention was not offered for sale or put into use in a public forum more than one year prior to filing for a patent in this country. Read about the on sale bar in the next section.

WHAT IS NOT PATENTABLE?

There are several factors that make an invention not patentable. They are:

- **Perpetual motion machines.** An inventor cannot receive a patent for perpetual motion devices, as the patent office does not consider them possible. If you've invented a machine that you believe falls under this category, call it something else, such as a "high efficiency energy generator."

- **Abstract ideas.** You must turn it into a legitimate invention and show that it works the way you say it works. In essence, your patent application will do this.

- **Laws of nature or naturally occurring substances.** For example, you can't patent a cheese from cows just because they are in a different region or climate and thus produce a new, unique flavor. The taste of cheese in all regions varies since it is affected by the grasses and grains a cow ranges on.

- **Inventions that are considered not useful.** Your invention must have some form of valuable use. For example, you can't patent square tires unless you can prove there are some real benefits. (Then again, maybe you can patent square tires if you can prove they would be of value as an anti-theft device. It would be difficult to drive the car away!)

- **Public disclosure for United States patent.** An inventor cannot receive a United States patent for an invention they have publicly disclosed more than twelve months prior to filing. Public disclosure includes any sale or offering for sale, exhibit at trade show, or being printed in a publication, with a few exceptions.

- **Public disclosure for foreign patent.** An inventor cannot receive a foreign patent if at any time the invention was disclosed publicly prior to filing a patent application. If you plan on receiving international patent protection, discuss the disclosure laws with your patent attorney before making a public disclosure.

You should seek a patent attorney's opinion if you have any question whether your invention is patentable. It should also be noted that you do not need a prototype when seeking patent protection—you only need to be able to describe the invention in sufficient detail so that one skilled in the art could construct your invention.

IMPORTANT LAWS

The United States has the most advantageous patent laws in the world for independent inventors, small companies, and corporations. Learn these laws and use them to your advantage.

Disclosure Requirements

United States law does not require inventors to understand how or why their invention works. Also, inventors do not need a prototype of it to receive a patent.

However, inventors must be able to describe their invention in sufficient detail so that *one skilled in the art* of their technology may construct the invention from the patent disclosure without undue experimentation (35 U.S.C. §112). In other words, you should prepare your patent application using terminology you would expect to be used by one skilled in the art of your invention. From this perspective it's important to understand how your invention works, what the inventive matter is, and how you can describe it in broad terminology.

ONE SKILLED IN THE ART

A person skilled in the art of your invention is not necessarily an expert in the field of your technology—it is merely a person with average knowledge in the industry.

Applying First-to-Invent Laws

The United States is a "first to invent" country—not "first to file." This means that only the first, true inventor(s) will be acknowledged as the patent grantee(s). An invention or discovery that has been diligently pursued by an inventor and has not been abandoned, has precedence over subsequent discoveries which are the same or similar in scope.

If there is confusion between two or more persons who have filed patent applications or have received patents on the same subject matter, the inventor who can prove his or her discovery has earlier priority over the other, will have the valid patent. This is regardless of who filed first or which patent was granted first. From this perspective, it's important to maintain excellent records in your journals.

During the prosecution (while your patent application is pending), the revelation of such a matter constitutes what's referred to as *interference.* It is very rare, but it can happen. One of the outcomes of writing a patent as explained in *The Patent Writer* is to improve your odds in the event this does happen. You'll learn how to write broader, more effective patent applications that may diminish the potential for interference.

For example, if there is a potential interference problem, but your patent application is broader than the other one, yours may cover subject matter that is not only not included in the conflicting patent, but is clearly dominant over it. This may be important in such a proceeding as a broader declaration of use, for example a method of use that is not claimed in the other patent application, would not be in conflict. In other words, there is no interference on that subject matter. Therefore, even in the event that another patent was determined to have priority over yours, subject matter disclosed by your patent application may still earn a patent if that information was not included in the priority patent. Better yet, it may actually dominate the prior patent.

The On Sale and Public Use Bar Rule

This term is also commonly referred to as the "one-year rule." As described in the United States Code 35 U.S.C. §102(b), an inventor may file a United States patent application for an invention up to one year after it has been publicly disclosed. A first public disclosure most frequently is when an inventor or company offers an invention for sale or makes a sale. But it can also occur through any number of other means such as publishing in a trade journal and so on. However, this rule does not apply to the mere attempt to sell or license only the patent rights. There are exceptions to the one-year rule such as experimentation, so you should consult with your patent attorney after any potential public disclosures.

TIMELY FILING

First-to-invent rights may not affect how you write a patent application, but you should be aware of this important law as it may influence when you file your applications. Also, since there is proposed legislation that could change the present law, make sure to visit www.patentwriter.com for legislation updates.

ONE-YEAR RULE

If you have a potential public use (i.e. sale, offer for sale, public disclosure), mark it on your calendar and then calculate one year from that date. You will need to file your patent application at least before the one-year anniversary of this date. As always, seek the advice of a qualified patent attorney if you believe you might have a potential public disclosure.

United States law does not prevent an inventor from filing a patent application for *improvements* made after the public disclosure or for subject matter not disclosed in the public disclosure. Frequently new improvements are made after the product has been launched and at times, even made in a public environment. This is particularly true with certain commercial and industrial innovations, where the inventor may initiate an improvement on the spot in order to overcome a certain user objection.

So, after the one year period, an inventor may file a patent application on a certain improvement, whether that is the product itself, a related manufacturing process, or how the innovation is used.

Foreign Countries—Absolute Novelty

In most foreign countries, with only a few exceptions, any public disclosure prior to filing a patent application prevents an inventor from receiving a patent in the foreign country. The one-year-on-sale bar does not apply to foreign applications. The information relating to a public disclosure may not be readily available to the foreign patent examiner determining patentability. However, the information may be revealed at a later date, possibly years later, during discovery when attempts are made to enforce the foreign patent against an infringer through litigation.

The threat created by the absolute novelty requirement means that great care must be exercised to avoid losing valuable foreign patent rights. Each of the following acts, without a confidentiality agreement or filing a patent application first, may destroy novelty:

- **Outsourcing.** Hiring outside tool and die makers, outside equipment manufacturers, or other outside consultants.

- **Sample Testing.** Supplying samples for testing to third parties who could evaluate the samples commercially or field test the invention.

- **Lectures/Papers.** Presenting details of the invention in professional lectures and papers in a professional journal.

- **Joint Ventures.** Openly divulging the invention in business dealings.

- **Visitors.** Permitting visitors to view installations and procedures.

- **Idea Submissions.** Submission of ideas to another on a non-confidential basis.

Should you have any questions about preserving your worldwide filing rights, consult your attorney.

WORLDWIDE FILING RIGHTS

If you want foreign patent protection, it is important to keep in mind that you need to file the foreign patent application within one year of your first United States patent application in order to claim priority. For design patents, you must typically file the foreign patent application within six months of the first United States patent application.

Foreign patent protection can be relatively expensive, ranging from $3,000 to $7,500 per country. To make matters worse, many foreign countries do not have adequate laws to protect the owners of patent rights.

If you are considering filing for foreign patent protection, the Patent Cooperation Treaty (PCT) is a recommended approach. The basic strategies you may use are discussed in Chapter 11. If you only want patent protection in European countries, you can file a European patent application. The cost for a European patent that presently includes over twenty European countries generally runs between $30,000 and $50,000.

Regardless of which option you choose, you will want to use your patent attorney, who typically has associates in foreign countries to assist in the filing of foreign patents.

INVENTORSHIP

An inventor is any individual who makes a contribution to at least one claim of the patent application. Please note that an inventor does not have to contribute to every claim. If two or more individuals make an invention jointly they have to apply for a patent as *joint inventors.*

An individual or entity that only makes a financial contribution is not a joint inventor and cannot be joined in the application as an inventor. Nor is someone who is following your instructions to engineer, make, or prototype your invention. Last, anyone who is involved in the manufacture of an invention who applies "best-way methodologies" would also not be considered an inventor. For instance, a mold maker who uses known production processes and tricks of the trade to improve production output would not be considered an inventor.

Upon filing your permanent patent application you'll sign an *Oath of Inventorship,* a declaration verifying that you believe you are the first, true

---

PRESERVING WORLDWIDE RIGHTS

Worldwide filing rights may be preserved through the filing of a provisional patent application prior to public disclosure in the United States. This is one of the key benefits to writing and filing inexpensive provisional patent applications.

JOINT INVENTOR WARNING

All co-inventors are equal owners of the patent rights unless a written agreement is executed between the co-inventors. Any co-inventor can license or sell his or her rights to a third party without paying anything to the other co-inventors.

inventor. An oath of inventorship is not required with the filing of a provisional patent application.

WHO CAN FILE?

By federal statute (35 U.S.C. §111), only the *inventor(s)* may apply for a United States patent, with only a few exceptions. Any patents that are issued from a patent application filed in the name of a non-inventor may potentially be invalidated. A person who falsely claims to be the inventor can also be subject to criminal penalties.

There are three statutory exceptions to the general rule that only the inventor(s) may file a patent application. The three statutory exceptions are:

1. **Refuses to sign or is missing.** If a joint inventor refuses to join in an application for patent or cannot be found after diligent effort, the application may be made by the other inventor on behalf of himself or herself and the omitted inventor (35 U.S.C. §116).

2. **Deceased.** If the inventor is deceased, the patent application may be filed by the legal representative of the estate (35 U.S.C. §117).

3. **Proprietary interest.** If the inventor refuses to file a patent application or cannot be found, a joint inventor or another person having a *proprietary interest* in the invention may apply on behalf of the non-signing inventor (35 U.S.C. §118).

PATENT OWNERSHIP

Patents have the attributes of *personal property* and may be sold, licensed, mortgaged, bequeathed by will or passed to the heirs of a deceased patentee. Ownership of patentable subject matter is determined under state law, not federal law.

Inventors

The presumptive owner of a patentable invention is the inventor. If there is more than one inventor, the inventors own the patent rights *jointly* unless otherwise agreed in writing. However, the inventor may transfer his or her ownership interests in the patent by a written assignment agreement to a third party or a co-inventor.

If two or more inventors jointly hold a patent, each individual may practice the teachings, without the consent of the other co-inventors. It's important to determine ownership rights beforehand to avoid precarious situations. For example, one co-inventor may use the technology in his own business, whereas a second co-inventor may choose to license the patent to his competitors.

Employment Agreements

All individual inventors should be aware of potential ownership issues with their employers. Many employees sign employment agreements that specifically transfer ownership of an individual's intellectual property rights to the company.

You should thoroughly review any employment agreements you have with your employer during the initial stages of the invention process.

You should also be aware that in most states it is illegal for a company to claim ownership of an employee's invention that is considered outside the business scope of the employer, and which was developed without any use of the employer's equipment, facility, or other assets of the employer.

For example, a software company cannot claim ownership of an employee's invention of a plastic boat ladder if the inventor did not use any company equipment, time, or other company assets during the development of the invention.

However, even when an agreement is not in place and an invention was created or perfected through using some of the company's assets by an employee, the company may potentially use the invention under a royalty free, non-exclusive license. This is called "shop rights."

Hired to Invent

Even if you do not have an employment agreement, your employer may still be able to acquire ownership of your patent rights if you were "hired to invent." Engineers, scientists, and product designers are all individuals subject to the hired to invent rule, also known as "work for hire."

Your state's laws determine whether or not ownership of your patent rights will be transferred to your employer. If you are concerned about the ownership of your patent rights, you should talk to a patent or business attorney in your state that understands employer-employee patent ownership.

TIP FOR EMPLOYEES

If you are an employee for a company, regardless of your position, review all legal documents you signed with your employer and the employee handbook. Some companies have broad clauses or procedures regarding the ownership of inventions created by employees. Don't ignore this. It is better to address any potential ownership problems with your employer before spending significant amounts of time developing your invention. If there are potential ownership problems with your employer, you should immediately consult with your attorney.

MAINTENANCE FEES

All utility patents which issue from applications filed on and after December 12, 1980 are subject to the payment of maintenance fees, which must be paid to maintain the patent in force. These fees are due at 3.5, 7.5, and 11.5 years from the date the patent is granted. Make sure to visit www.patentwriter.com for additional information about calculation of these dates and the amount of the fees.

A six-month grace period is provided after the due date where the maintenance fee may be paid plus a surcharge. The USPTO does not mail notices to the patent owners to notify them when the fees are due. Failure to pay a maintenance fee on time may result in expiration, or abandonment, of the patent. It should be noted that the maintenance fees are only required for utility patents—not design or plant patents.

PATENT COSTS AND FEES

There are government filing fees and costs associated with patenting. You'll want to understand them, as they can affect your patent-drafting strategy. For example, filing a provisional patent application initially costs much less than filing the permanent patent application.

Small Entity and Large Entity Status

Fees are based on the size of the company. Small entities are those companies that have a maximum of 500 employees, with large entities being over 500.

If a patent is licensed by a small entity to a large entity then it must pay the large entity fees. In such a case, the change from small entity to large entity occurs once a commitment has been made by the smaller entity to license to the larger one.

With the exception of the patent search report fee, small entity fees are one-half that of large entities.

U.S. Patent Office Fees

The U.S. Patent Office has several fees you should be aware of when filing a patent application. The costs for 2006 for the more common actions, based upon *small entity status* (this includes inventors), are listed below:

U.S. Patent Office Fees	
DESCRIPTION	**FEE**
Filing Fees	
Provisional application filing fee	$100
Permanent utility patent application filing fee*	$500
Each independent claim in excess of three	$100
Each claim in excess of twenty total claims	$25
Design patent application filing fee**	$215
Plant patent application filing fee***	$330
Issue Fees	
Utility patent issue fee	$700
Design patent issue fee	$400
Plant patent issue fee	$550
Maintenance Fees	
3.5 year maintenance fee	$450
7.5 year maintenance fee	$1,150
11.5 year maintenance fee	$1,900
*This utility filing fee includes the $150 basic fee, the $100 examination fee, and the $250 search fee.	
**This design filing fee includes the $100 basic fee, the $65 examination fee, and the $50 search fee.	
***The plant filing fee includes the $100 basic fee, the $80 examination fee, and the $150 search fee.	

Patent Office fees change frequently—often times yearly (usually October of each year). For the most current U.S. Patent Office fees, visit www.patentwriter.com, which is updated to reflect the most current fees you need to be aware of. Your patent attorney will also be on top of current and proposed patent related fees and should notify you in advance so you may take action prior to the effective date, should you wish.

As you can see, a hefty filing fee can result if you don't control the number of claims in the application. By following *The Patent Writer's* format you can avoid excessive fees, albeit at times, it may be worth paying for additional claims if it improves the scope of the patent. Some patent applications may include a substantial amount of patentable subject matter, making it worth the investment.

An option to filing one patent with many claims may be to file additional patent applications more narrowly defining the inventive material. Generally speaking, a well-written application will contain about two to three independent claims and about ten to twenty total claims, but these numbers can vary widely—some patents may have more than two hundred claims.

OTHER RELATED COSTS

MONEY WELL SPENT

If you have an invention potentially worth millions of dollars, spending $5,000–$20,000 will be a very smart investment in protecting your invention.

In addition to the above patent office fees, there are additional costs. These costs depend upon how much you and your company's internal engineering and legal department may or may not contribute to the patent-drafting process. Additional costs are:

Attorney's Fees

These fees are in addition to any U.S. Patent Office filing and issuance fees, and will depend upon four primary factors. The first is the attorney's hourly rate and second is the complexity of the patent application. A simple, short provisional patent application would take far less time to write than an in-depth, multi-invention application with dozens of claims.

The third determining factor is based upon how much of the patent application you've written and how well you've written it. Following *The Patent Writer's* instructions will undoubtedly help you save hundreds, perhaps even thousands of dollars. There will be a significant reduction in legal fees an attorney will charge if he or she is only reviewing and editing a provisional patent application, or if he or she is hired to add the legal claims to a permanent application.

The fourth is more difficult to plan for—the prosecution costs. While a patent is pending and being prosecuted, there could be considerable communication between the patent attorney and the patent examiner. Typically your patent attorney will argue on your behalf to get the patent office to approve and allow the broadest possible interpretation of the claims in your application. Your patent attorney may also need to argue on your behalf in the rare case of an interference proceeding.

Since there is no telling what issues may arise during prosecution, it is nearly impossible to calculate your projected costs. However, at least planning and budgeting for prosecution costs is a smart practice. Ask your patent attorney for an estimate of the costs associated with prosecuting a patent application with subject matter like yours.

Drawings

You may incur some charges for the drawings in your patent application. Even if the first drawings submitted to the patent office are not top quality, you'll have to have them formalized (perfected) before the patent issues. It could cost a couple hundred dollars for simple drawings or over a thousand dollars for complex drawings. In Chapter 9, Drawings, you'll learn what's required, how much you may be able to do yourself, and how you may be able to defray some of those costs.

Test Analysis

At times, you may be burdened with the proof to have certain inventive subject matter verified by a laboratory. Keep in mind that if an invention is novel and has never before been used, then it may require some scientific substantiation to help qualify the invention. For example, if you are trying to illustrate how a new folded structure using a thin gauge plastic material maintains its rigidity, you may have to include a lab test to prove the point. Scientific tests of this type are included in only about 5 percent to 10 percent of patent applications, depending upon the field. When you conduct a patent search, you'll find some examples.

Mailing and Postage

Nominal fees for shipping or postage will be required, but generally speaking this should not exceed $100 over the course of filing the patent application through the prosecution phase.

Now that you know about the key parts of a patent, you've estimated your costs, and you have a good understanding of the patent laws and the overall process of filing your patent, you're *almost* ready to get down to the business of writing your patent. Before sharpening your pencil, you must first determine *how* you are going to write your patent. You have a little more homework to finish in order to clearly establish your objectives and effectively plan your patent-writing approach. Planning is critical, and without designing your roadmap to patent writing success, you're probably not going to arrive at your destination.

Chapter 3

Patent Objectives

Protect, preserve, and profit.

Y ou can say that your patenting objective is to protect your invention, preserve your future, and profit from its sales. This is not going to happen with just any patent, however.

It is commonly stated that more than 95 percent of patents never make money for the inventor. Separate yourself from the 95-percent-and-over failure category and take the necessary steps to be successful. It all starts with a little background on patent drafting and product development methodologies, and you'll be on your way to maximizing your patent protection.

MAXIMIZING PATENT PROTECTION

It goes without saying that you'd want maximum patent protection from your patent applications. But make no mistake about it, to achieve this you better know what you're doing.

Are you aware of the various types of patent applications? Do you understand the various types of patent protection you can achieve? Have you thoroughly researched the marketability and novelty of your invention? Do you know the correct terminology used in your field? What goes into a patent application? How and when should you use your patent attorney?

WHEN TO SELF-DRAFT

You need to know when to self-draft and when not to self-draft. Following a hard-line practice where you always self-draft for all inventions can be a big mistake, just as hiring a patent attorney for all inventions can be.

43

If you can't answer these questions now, there's more to learn. The answers are essential to writing strong patent applications and maximizing your protection.

Provisional Application First

A provisional patent application is a valuable tool for inventors to start the patent process. Provisional applications are simpler to prepare for an inventor and have a significantly lower USPTO filing fee.

However, a provisional application still has the same disclosure requirements of a permanent application, except for the claims section. Finally, provisional applications are not examined by a USPTO examiner for patentability. To protect the subject matter in a provisional application, you must file a permanent application within one year.

Role of Product Development

As we've already said in earlier chapters, inventors often times make numerous improvements to their invention during product development. Provisional applications can assist you in filing for these improvements since any number of provisional applications can be combined into a single permanent application.

Many inventors (and businesses) file consecutive provisional applications on the same invention with each provisional application containing additional subject matter not contained within a prior application. You'll learn more about the use of provisional applications in the following chapter.

Patent Searching

As we have said a number of times already, if your invention already exists you cannot receive a patent on it. This means quite simply, that if anyone, in any country, at any time received a patent on the same invention that you now want to patent, then you will not be able to receive a patent.

Moreover, it's just as important to know that your invention has not been disclosed in non-patent literature such as technical or trade journals. If a college professor, for example, wrote a technical paper describing

---

AN IMPORTANT CASE

Applying the tenets and patent drafting terminology commonly used prior to the *Phillips v. AWH* case (decided July 2005) may have serious consequences on the scope of your patent. Your objective is to secure strong patent protection. Our objective is to show you how.

MORE THAN PATENT SEARCHING

Inventors should not limit their searching to just patents. There are many resources readily available that you can search for similar products such as catalogs, stores, books, technical journals, and the Internet.

what you are considering patenting yourself, your patent could be later invalidated, making all of your effort and investment worthless.

Searching Prior Art

With more than 7 million United States patents, and more than 35 million patents worldwide, coming up with a patentable invention is not always easy. Therefore, before you make a substantial investment in your invention, you'll first need to know whether anyone else has a patent on it or if it's already in use. This is referred to as *prior art.*

For this reason, searching for prior patents, products, and public disclosures is a critical step that every inventor should take *before* preparing a patent application.

First, there is no need to prepare a patent application for an invention that is not patentable. Second, to adequately prepare a patent application you need to know what is potentially patentable within your invention so you can focus upon the important subject matter, thereby strengthening your patent rights. In other words, if the product is already patented, but you come up with *improvements,* your patent search will reveal whether your improvements, or more accurately what improvements, may be patentable. (See Illuminated Hammer.)

Now, more than ever, it's important to conduct a preliminary patent search yourself to get a quick understanding of the landscape of technology in your field of invention. If you find other patents that obviously pre-empt your invention, then you have saved the cost of a third party professional patent search.

There are many online patent databases that you can search yourself, including public databases (www.uspto.gov for United States Patent Office, and www.european-patent-office.org/espacenet/ for the European Patent Office).

There are also numerous commercial patent databases that provide more sophisticated search tools than the government databases, including the ability to enter one search query, and simultaneously search patents from up to ten, twenty, or even thirty different countries at once. See the Resource List on page 225 for detailed information on search technology tools.

Regarding products, you should search catalogs, stores, and other materials where similar products may have been promoted. Finally, Inter-

ILLUMINATED HAMMER

Related patents found during the preliminary patent search:
1,234,567; 2,345,678; 3,454,343; 2,334,535; 6,345,634; 3,303,605; 1,234,567; 2,345,678; 3,454,343; 2,334,535; 6,345,634; 3,303,605; 1,234,567; 2,345,678; 3,454,343; 2,334,535; 6,345,634; 3,303,605. These patents revealed illuminated tools and other devices, various embodiments of hammer devices, but did not reveal the combination of "illumination" and "hammer" as envisioned by the authors in the present invention.

PATENTABILITY V. INFRINGEMENT

Many inventors incorrectly believe that if their invention is patentable they are not infringing upon another patent. Even if you receive a positive patentability opinion, you still need to consider all potential infringement issues.

net search engines are valuable non-patent literature resources in determining the patentability of your invention. See the Resource List for examples of some search engines.

After you have completed your patentability research, you should keep any material information you've uncovered in your journal or in files so it is not lost. You will need to disclose to the USPTO any patents that materially relate to the patentability of your invention, including any subject matter that may negatively affect the patentability of your invention.

Use your inventor's journal during your search and be sure to record comments and notes relating to discoveries that you make. The patent searching data you've collected will be useful when you write your patent application.

Learning Industry Terminology

Many inventors do not learn the terminology utilized in the industry of their invention, which can negatively affect the ultimate quality or strength of a patent. It is important for you to research the terminology utilized within prior art patents, industry publications, catalogs, and websites. Most of the terminology used in an industry may be discovered through patent searching.

You will want to utilize similar terminology within your patent application, so write down important terms that are difficult to remember. Remember, your patent is interpreted as one skilled in the art of your invention would interpret your wording.

To help you quickly learn industry terminology, PatentCafe's ICO Global Patent Search at www.patentcafe.com actually creates a terminology thesaurus in real time, specifically for the patents that relate to your field of invention.

Modeling Existing Patents

There is no need to reinvent the English language, simply use existing terminology and writing styles when you can. There are no copyright laws governing patents. Hence, feel free to copy words, phrases, and even portions of prior patents.

You will be surprised how the simple concept of modeling will improve your ability to write patent applications. But be cautious about a few points. First, model your patent applications after those patents that are

---

INTERNET

Take advantage of this free wealth of information. You can perform quick searches for technology that relates to your invention for the purpose of patentability, infringement, and marketability.

USE COMMON WORDS

Since the *Phillips* court case, it is more important to use terms in a manner that is consistent with their ordinary meaning to one of ordinary skill in the art of your invention. It is recommended that you read the most recent patents on technologies related to your invention, which will most likely utilize terms as used by those skilled in the art.

Do not be afraid to use terms in other patents—as long as they provide the meaning you intend!

---

well written and have been prosecuted by competent patent attorneys. In other words, just because someone has written a simple two-page patent and you like how it reads, doesn't mean it is a well-written patent with broad scope.

Also, when modeling after existing patents, look for those more recently issued patents. You don't want to use archaic terminology and patent-drafting structures that may have changed over the years.

USING A PATENT ATTORNEY

You will be hard-pressed to find a successful inventor that does not rely on his or her patent attorney for legal matters related to the invention. You will, however, be able to find thousands, even tens of thousands of "patent it yourself" inventors who have not achieved success.

Using a patent attorney does not guarantee success, but it does help ensure that you receive a strong patent in the first place, and that you have a staunch legal advocate if you find yourself in a situation such as patent infringement. The authors believe that a patent attorney should have a part in every new patent filed!

Patent Attorney's Role

Your patent attorney serves one main purpose: to advise you on how to adequately protect your intellectual property assets, and help you enforce your intellectual property rights. You will rely upon your patent attorney for patentability advice.

Your patent attorney will also be utilized to review your self-drafted provisional patent applications and write your permanent patent applications. Your patent attorney may also be utilized to determine whether you may be infringing upon a third party's patent rights.

After your patent is granted, you will rely upon your patent attorney for enforcement of your patent rights and maintenance of patent fees. All this may sound expensive, but the fact is, it's really not. Legal expenses are an inherent part of any business. When your patented products are generating sales and profits, the legal fees become a minor expense and are easily justified.

If you or your company delves into the world of patents with the mindset of "how are we going to afford the legal fees," then you're not prepared for success. You're clearly not thinking in terms of generating

ILLUMINATED HAMMER

Here is a short list of words used in the hammer and lighting industry—terminology that was not necessarily known to the inventors prior to conducting their patent search for the illuminated hammer. *DynamicSearchThesaurus*™ *DST:* Light, illuminate, illumination, lights, emitting, luminous, lighting, emits, emit, emitted, head, handle, grip, ergonomically, and ergonomic. (DST™ is a feature of PatentCafe's ICO Global Patent Search.)

BIG DECISION

Hiring your patent attorney is one of the biggest decisions you will make in the invention process. Your patent attorney will provide you with patentability advice, prepare your permanent application, provide infringement advice, and various other types of legal services. Since you will be using your patent attorney for many years after the filing of your application, make sure to pick a patent attorney that is right for you. Switching patent attorneys during the patent process is not easy and can be costly.

sales and profits and protecting them with patents. If this sounds like you, you should do more research and make sure that you are ready to delve into the business world, and that you have properly budgeted for patent legal costs.

Selecting a Patent Attorney

Your patent attorney is similar to a business partner—you will be relying upon the person for legal assistance to ensure adequate protection of your invention. Expect this relationship to exist for many years to come.

It is therefore important to make sure you take your time and adequately research who is the best patent attorney for you. In other words, you do not need the "best patent attorney" in the world; you need to find the patent attorney who best fits with you and your invention. Similarly, you do not want to choose the cheapest patent attorney, nor the most expensive one.

The following are some suggested factors to consider when selecting a patent attorney.

Experience

How much experience does the patent attorney have in preparing patent applications? Do not focus so much upon the number of patents received, but rather the quality of work performed.

Engineering Background

All patent attorneys not only have a law degree, they must also have a scientific degree such as mechanical engineering, electrical engineering, or chemical engineering. It is important that the patent attorney working upon your invention has an engineering background related to the field of your invention. If you have an invention dealing with semiconductor technology, you obviously will not want to hire a patent attorney with a mechanical or chemical engineering background! However, if your invention relates to a simple mechanical technology, most experienced patent attorneys should be qualified to assist you.

Trust

It is important that you trust the patent attorney you are working with. If you do not feel comfortable when you talk to him or her, then you should consider a different patent attorney.

Right Fit

Make sure the attorney's law firm wants you as their client. Some law firms specialize in working with large corporations and find it difficult to satisfy the needs of an individual inventor.

Scope of Services

Some patent attorneys specialize in patent prosecution (patent drafting and filing), but do not provide litigation services. Other patent attorneys only litigate and do not perform patent drafting or prosecution services. Still, other attorneys in small- to medium-sized patent law firms have attorneys on staff who can address the full spectrum of services likely to be requested by inventors throughout their invention career. Discuss with your attorney how he or she will provide the full scope of services if and when they are required.

Fees

It is important to determine the legal fees the patent attorney charges. Does he or she charge flat fees or hourly? When selecting a patent attorney, you are well advised to not choose the highest priced attorney, nor the lowest priced attorney. The best "bang for your buck" typically comes from middle to upper-middle priced attorneys. Also, choosing an attorney based solely upon the fee schedule is the wrong thing to do. What may seem like a bargain today may not be such a bargain if the attorney is unable to adequately protect your multi-million dollar invention!

Location

Many inventors feel a patent attorney should be located next door to them. However, with today's technology (e-mail, fax, priority couriers), inventors now have a larger market of patent attorneys to select from. You are no longer limited to the patent attorney down the street—you now have thousands of options throughout the United States!

TRADE SECRETS AS AN ALTERNATIVE

You should consider whether trade secret protection is an appropriate

LOCATION

With the Internet, it is very easy to locate and communicate with quality patent attorneys throughout the United States. The Internet has given inventors with a previously limited selection of patent attorneys a buffet of choices.

 ILLUMINATED HAMMER

Commercialization Objectives:
Find a licensee for the invention. Limit up-front cash investment by spending more time than money.
Patent search: $100 for do-it-yourself search on a commercial patent database.
Writing patent application: time, not money.
Attorney review of our provisional patent application: $500 to $1,000.
Filing fee: $100.
Marketing expenses to find and negotiate terms of a license agreement: $1,500 over twelve months.
Cost to have attorney refine the license agreement: $2,000 to $5,000—split 50/50 with the licensee.
License fee: $10,000 (we have all of our investment back at this point).
Future royalty of 3.5 percent of sales for twenty years.

ILLUMINATED HAMMER

The authors (also the inventors) ran an invention evaluation on the illuminated hammer illustrated in Chapter 1 that identified some shortcomings of the original design. The solutions to these problems became new patentable subject matter that was incorporated into the provisional patent application (PPA). Holding off initial filing so that we could include this additional matter in the PPA helped ensure stronger future patent protection—and the potential increase in return on our investment. For example, the impracticality of drilling holes in a hammerhead and the questionable means of securing a light bulb to the hammerhead itself raised serious questions regarding cost and durability. Thus, we were forced to improve the design.

alternative to patent protection. Generally speaking, trade secret protection is not a viable alternative to patent protection.

However, there are instances where meaningful and lasting protection may be received from trade secret protection and should be seriously considered (manufacturing processes, recipes, software technology, etc.).

It's important to understand the benefits, as well as the potential liability or risk, in choosing trade secret protection over patent protection. Be sure to explore the options, along with the associated advantages and disadvantages with your patent attorney.

COMMERCIALIZATION OBJECTIVES

The number one objective for seeking patent protection is to make money! While that sounds intuitive, the fact is, making a return on your patent investment means that you must make an investment of time and money in the first place; must choose what the best invention or technology is in which to make an investment; and must determine how that investment plays into your overall business plan. You'll need to balance all of these considerations, and more, in order to determine your objectives.

Cost and Budget Considerations

Famous author, philosopher, and actor Will Rogers once said, "It is not the return on my investment that I am concerned about; it is the return *of* my investment." Wise words, indeed. All inventors are continually challenged with balancing how much money they have, and where the money should be spent to ensure the greatest probability of financial success.

Unfortunately, the cost of a patent attorney (one of the most important investments) usually slips to the bottom of the must-pay list. You'll need to take a hard look at budgeting and costs to obtain a satisfactory return on your investment.

While investment in a patent attorney is important, it's also important to make sure that you don't over-invest in legal fees. Whether you represent a small entity company of one employee, or a company with 499 employees (still considered as a small entity), the advice is the same: keep patent legal fees under control and do as much of the advance work as possible yourself. This includes writing your first patent draft, performing your own preliminary patent search, and matching the right attorney for the job.

You'll also need to craft a budget that reasonably represents the amount of money that you are committing to developing, filing, and protecting your intellectual property. Every responsible company works under a budget. The investment in a new invention must include how much is going to be invested in prototyping, market testing, patent legal fees, and if you are planning to make and sell your own invention, a budget for manufacturing and sales. Once you have the budget, stick to it!

If you are pursuing your invention without a good understanding of the potential profit, without a plan and a budget, then you're probably wasting your valuable resources.

Companies are also being challenged to decide which innovations to pursue and patent, and which innovations to either abandon or defensively publish. Because you will need to make similar choices, it's important that you perform your marketability analysis/invention evaluation to help you determine which of your inventions are most likely to succeed.

Other budget considerations extend to the revenue side of the equation: will you expect to generate monthly cash flow from your inventions through licensing royalties, or do you expect to generate sales revenue, keeping the patents in force to protect your market share?

Finally, if you plan on filing for international patent protection, you will need to get together with your attorney and run various cost scenarios in those countries where you expect to generate licensing or sales income. The same principles outlined above will apply to your patent investment assessment.

Regardless of what budget decisions are made, they should all support and reinforce your commercialization objectives. Smart investments in your intellectual property should all point to you making money!

ROLES OF A COMPANY'S DEPARTMENTS

Whether your company is a one-man-show or an organization with 500 employees, attention must be paid to the many functional aspects of business.

This means that every new invention will impact, and will be impacted by, decisions based on marketing, manufacturing, finance (budgets), engineering, and research and development (R&D).

How do these functional areas of business affect your patent strategy, patent investment, and most importantly patent success? Let's take a quick look.

UNWISE TO PAY TOO LITTLE?

"It is unwise to pay too much, but it is worse to pay too little. When you pay too much, you lose a little money—that is all. When you pay too little, you sometimes lose everything because the thing you bought was incapable of doing the thing it was bought to do. The common law of business balance prohibits paying a little and getting a lot—it cannot be done. If you deal with the lowest bidder, it is well to add something for the risk you run, and if you do that you will have enough to pay for something better."
—John Ruskin (1819–1900)

51

Marketing Department

The marketing department plays two key roles in patent development. First, it helps identify new products, technologies, or product improvements that can be marketed and sold for additional profits. The marketing department is also responsible for market research and marketability analysis of proposed new inventions. Yes—even large companies with marketing departments must go through the same invention evaluation process that independent inventors must complete. Of course, these improvements may later be patented to protect the new revenue stream.

Second, the marketing department will be responsible for taking the new inventions to market. Once the new products are on the market, the marketing department will be the front line for customer feedback, yielding even more future improvements and innovation.

If you or your company plans to license your inventions and patents, you must make sure the innovations appeal to the marketing department of potential licensees and that your patents adequately protect their interests. From this perspective, you'd be wise to have marketers on your team as early as possible to help you qualify the marketability of the inventive subject matter.

Manufacturing Department

The manufacturing department must ensure that a new invention can be manufactured cost-effectively. We mentioned earlier that even if new products offer significantly more features and benefits, the sales group would only be able to generate revenue if the products hit the market with a competitive price tag.

Therefore, just as independent inventors must perform a manufacturability assessment, so too is the manufacturing department charged with ensuring that proper manufacturing methods are used to create a cost-competitive product.

Finally, there's another role the manufacturing department can play in the intellectual property arena. If novel manufacturing methods need to be developed in order to produce competitive products, these methods may be eligible for patent protection themselves. This added patent protection improves the intellectual property (IP) value of the invention.

If you discover new manufacturing methods that you believe give a competitive advantage, write them down in your invention journal first, then get together with your patent attorney to discuss the possible next steps to protecting those methods.

Remember, in the overall picture, the investment in manufacturing process or methods patents must satisfy the business objectives—they must be solidly tied to the company's ability to make money.

Engineering Department

Without question, the engineering department is where inventions can get "reduced to practice." This is frequently where product development occurs. We won't go into the details of what the engineering department does with respect to patent development. We do want to point out, however, dangers that engineering departments can bring into the patent picture.

First, some engineers may get carried away with developing inventions. Just like mountain climbers like to climb Mt. Everest "because it's there," engineers may create inventions "because they can." For example, we've seen engineers design special headed bolts for specific automotive applications. While the bolts served a purpose, traditional bolts would have done just as well. Further, there was no market for the bolts outside of their limited application, so there was no way to generate additional competitive advantage or sales from the engineering or patent investment.

Keep the engineering tasks focused on the objectives, and evaluate the marketability or return on investment (ROI) potential, before investing. You cannot blindly pursue and patent every invention engineering brings forth.

Finance Department

The finance manager controls the purse strings. In the micro-business, this could be your husband or wife. Either way, the financial manager should press the other departments for their budget projections, perform the financial analysis, and determine where the money will be spent in order to produce the greatest return (or lowest risk) on that investment.

The financial manager should also keep a sharp eye on legal fees. As legal costs can get out of control, make sure that invoices specify the scope of legal work performed. Check with engineering or other managers to

EDUCATE EMPLOYEES

Many companies do not take the time to educate their employees about intellectual property and what the company expects from them. It is often times helpful to bring a patent attorney into the company for a seminar and discussion relating to the company's intellectual property rights.

ensure that that scope of work was necessary. Occasionally check invoices against past invoices to see if any changes in the attorney's billing process or billing rates have occurred.

The finance managers in large corporations play other roles including patent portfolio analysis, intellectual property acquisition analysis, license revenue forecasting, and other finance and accounting functions. While important, these other responsibilities are not within the scope of *The Patent Writer*.

Other Company Considerations

It's important to address myriad other intellectual property issues in every company. These may include policies related to the use of nondisclosure agreements (NDAs), also referred to as confidentiality agreements, management of trade secrets, invention assignment agreements for employees and outside contractors, and so on. Your patent attorney can help you review your company's policies and practices and help you put effective controls in place.

It may be difficult to calculate how these additional considerations will produce a return on the investment, but it is relatively easy to understand that if an employee inadvertently lets out a trade secret, the results could jeopardize the company's very survival. Thus, it is essential that every company, large or small, has an employee trade secret and IP program in place.

Address these other considerations, but keep a reasonable perspective on the investment you will make in legal fees to bring your intellectual property plan up to date.

It is important to identify your "objectives" up front *before* you start writing your patent application. Your objectives will define not only how you write your patent application, but how you position your invention to be the next great thing!

Provisional Applications

The inventor-entrepreneur's starting point—provisional patent applications—cater to those that are fast, flexible, and focused.

Provisional patent applications, also referred to as PPAs, are commonly used in industry in virtually every field and every size company. There are good reasons why. They are relatively inexpensive, can be written relatively fast, and when properly used, they establish a legal priority date and preserve your worldwide filing rights.

Provisional patent applications are a perfect tool for most small businesses and product developers to use to protect their innovations during the initial development phase. Instead of running up legal fees on costly permanent patent applications, a small business can smartly use that money to qualify product marketability instead.

It's really just common business sense to use provisional patent applications whenever you can to cut costs, preserve your intellectual property rights, and leverage your investment across many inventions, instead of over-investing in a single invention that may or may not pay off.

In this chapter, we'll show you the advantages, as well as disadvantages, associated with the use of a PPA so you may become proficient in its use, should you decide to do so. Frankly, for almost all small businesses, filing a provisional patent application before filing a later permanent application is probably the best first step to take.

GREAT TOOL FOR INVENTORS

Provisional applications are a great tool for inventors during the early stages of the invention process where they are attempting to determine the marketability of their invention. However, make sure to educate yourself about the patent process before filing a provisional application so you understand all of the benefits and detriments.

OVERVIEW

A *provisional* application is a United States national application for a patent filed with the USPTO under 35 U.S.C. §111(b). Provisional patent applications do not require the inclusion of formal patent claims, oath/declaration of inventorship, the abstract, or prior art disclosure. It provides the means to establish an early effective filing date of a patent application and allows the phrase "Patent Pending" to be applied to goods.

Simply put, a provisional patent application is a simplified version of the permanent patent application. The more costly parts of preparing the permanent patent application are in the filing fees, search report fees, and professional legal fees associated with writing the claims. Since claims are not included in the provisional application, the legal cost to prepare it is significantly lower. Likewise, the parts of a provisional patent application are much easier for a layperson, engineer, inventor, or designer to write.

This is one of the primary objectives of *The Patent Writer*—that is, showing you exactly how you can write a successful provisional patent application. Then you can have your patent attorney review and edit your draft.

Many expert inventors and engineers have become proficient at writing and filing provisional patent applications themselves. This is something you and your patent attorney may consider as you become more proficient.

A provisional application provides patent pending status for only one year. A provisional application *automatically expires* at the end of this one-year period and cannot be extended.

A permanent patent application must be filed within this period to claim priority from the filing date of the provisional application. The period of up to one year of pendency for the provisional application is excluded from the term calculation of a granted patent that claims the earlier of the provisional application priority date. Consequently, a total patent term can be twenty-one years from the PPA filing date.

ADVANTAGES OF PROVISIONAL PATENT APPLICATIONS

There are several advantages to filing a provisional application. Below is a listing of benefits that are recognized by the USPTO:

- **Provides Simplified Filing.** You can publish notice that your inven-

SIGNIFICANT EFFORT REQUIRED

Some inventors make the mistake of just filing one to two pages of rough sketches and a little written description. Provisional applications require the same effort as a permanent application for them to be a valid supporting application. Failure to take the time to prepare a quality provisional application can be costly. If you are not going to make a significant effort in preparing your provisional application, the application you file may not be worth the $100 filing fee you pay.

MYTH

You may hear from people that a provisional application is not a "real patent application." This is incorrect. A provisional application has the same legal effect as a permanent patent application and should be treated with the same respect.

tion is patent pending immediately after filing your PPA, which can be a relatively short time frame compared to filing the permanent patent application.

- **Lower Initial Investment.** The filing fee for a provisional application and any related legal fees are far less than that of the permanent application.

- **Defers Patenting Costs.** By initially filing provisional patent applications instead of permanent ones, you will be deferring the overall costs of filing and prosecuting permanent applications by a year or so.

- **One Year Market Assessment.** You may use one full year to assess the invention's commercial potential before committing to the higher cost of filing and prosecuting a permanent application. If you decide to abandon or defensively publish the application, you're forfeiting a much smaller investment.

- **Establishes Priority Date.** Filing a provisional application establishes an official United States patent application filing date for the invention. Establishing a priority date becomes important if there is an interference proceeding at some later date. The inventor who has the earlier priority date is in the senior position and the burden of proof then becomes that of the other inventor, who is considered to be in the junior position.

- **Patent Pending Notice.** Permits authorization to use and print a "Patent Pending" notice in connection with the invention for one year. So, you may publish the notice on your products, literature, websites, and so on. Giving public notice early on can be a wise step in thwarting competitors.

- **Paris Convention Priority Year.** The Paris Convention provides the right of priority with regard to applications filed in foreign countries after a provisional application is initially filed in the United States. Provided an applicant files in the foreign countries within one year of the provisional application filing date, the provisional application date also becomes the effective filing date in the various foreign countries.

- **Greater Security.** Enables immediate commercial promotion of the invention with greater security against having the invention infringed. The security of having a U.S. patent pending goes a long way towards protecting your future rights.

DETERMINE MARKETABILITY ASAP

If you are filing your provisional application to determine whether your invention is worth paying a patent attorney to prepare a permanent application, you should start to determine the marketability of your invention in the marketplace immediately after filing the provisional application. Some inventors make the mistake of waiting six to nine months after the provisional filing date and must initiate the permanent application process before they are able to fully determine the marketability of their invention.

CANNOT ADD NEW
MATTER

As with a permanent application, new matter cannot be added to a provisional patent application. However, you can file another provisional patent application containing the new subject matter and then file a permanent application combining the multiple provisional applications.

- **Confidentiality.** Provides the same confidentiality, access, and certified copies by the USPTO as §111(a) permanent applications for patents. Provisional patent applications are not reviewed, nor are they publicly disclosed during the first year of pendency. Once a permanent patent application is filed claiming priority of the provisional application, the provisional application then becomes part of the file folder, and becomes subject to eventual public disclosure.

- **Filing Multiple Provisional Applications.** Allows for the filing of multiple provisional applications and for consolidating them in a single §111(a) permanent application. When we talk about provisional patent application strategies later in this chapter, you'll see how important this becomes.

Provisional patent applications are time-proven instruments and have withstood legal discretion. Should you use one in your invention activities, you can do so with confidence.

DISADVANTAGES OF PROVISIONAL PATENT APPLICATIONS

Even though there are many advantages to filing a provisional application for self-drafters, there are a few disadvantages that you should be aware of, such as:

MUST FILE WITHIN
ONE YEAR

You have to file both a permanent United States patent application and any foreign patent application within one year of your provisional application's filing date if you want to claim "priority" of your provisional application. Filing a provisional application too early may force you to file a permanent, foreign, or PCT application before you are fully prepared to do so.

- **One-Year Deadline.** You must file a permanent application within one year of the provisional application's filing date to maintain the priority date, which needs to be closely monitored. The best way to remember this important date is to have your patent attorney put the expiration date on the docket. Thus, you will automatically be notified prior to the expiration date so the permanent application can be filed.

- **Not Examined.** The provisional application is not examined by the USPTO; hence, you are effectively delaying your patent protection. This may become an issue if a competitor begins offering potentially infringing products. In such a case, you'd be wise to promptly file the permanent application and get it issued. Filing your permanent application will engage your patent protection. Meanwhile, you can let the competitor know that you have a patent pending and that you intend to either license it, or enforce your rights as the case may be.

- **Design Patent Applications.** Provisional patent applications cannot be used for priority purposes for later-filed design patent applications. If your invention can potentially be protected by design patent protection, you may want to file a design patent application simultaneously with the provisional patent application.

- **Foreign Rights.** If you want to file in a foreign country and want to utilize your United States filing date for priority purposes, you have to file the foreign patent application within one year of your first filed application that, in this case, would be your provisional application. Remember that you have only one year to file your international application or non-provisional U.S. application, but more importantly, you should give serious consideration to filing your provisional application with at least one claim. A provisional patent application without a claim could, in the future, be considered as an invalid priority application by the PCT.

Overall, the advantages of using provisional patent applications generally outweigh the disadvantages, providing they are used properly, and at the appropriate time in the innovation cycle. Read on and you'll be able to determine which strategy best fits your invention's development needs.

WHEN *NOT* TO USE A PROVISIONAL PATENT APPLICATION

There are times when it is best to file the permanent patent application first, instead of the provisional. Here are four of the more common reasons you may consider skipping the filing of a provisional patent application and filing a permanent patent application instead.

When filing the permanent application instead of the PPA, follow the same patent-drafting procedure we've outlined in the book. The only difference is that after you submit it to your patent attorney, he or she will add the legal claims and submit it to the patent office.

Marketability Is Established

If you have established that your invention is marketable (e.g. solid research, product sales, etc.), then you should considering filing a permanent patent application through your patent attorney. Make sure you are objective in your marketability determination and not subjective. With marketing underway, it is a lot easier to justify the cost of the permanent patent application.

Infringement Is Anticipated

If infringement is anticipated or perhaps considered imminent, time may be of the essence. Having an issued patent with the legal claims granted would be desirable, since patent infringement is not enforceable until a patent is issued.

In such a case, you'll want to pursue prompt filing and prosecution of the permanent patent application. Consult with your patent attorney on this strategy. There are some administrative steps they may take in order to expedite prosecution.

Rapidly Changing Technology

You may have developed a technology that is in a rapidly changing field. This is fairly common in the field of software development and computer design. To prevent others from using your technology, you'll want to have a patent issued as soon as possible.

There's one drawback to using this approach in the field of software, however. The software review unit at the U.S. Patent Office has been significantly backlogged over the past few years. The review of a patent application in some fields may take up to three years, maybe longer.

Funding a Start-up

Many investors will want to see patent protection secured prior to funding. Keep in mind that banks, venture capitalists (VCs), and angel investors are not experienced in the field of patenting and most likely not experienced in the field of your invention. They'll want to see the patent, or patents as the case may be, issued so they'll be able to qualify the scope.

As you know, a company's intangible asset value may be as great as 85 percent of its overall assets. Provisional patent applications essentially have little value, whereas a well-written issued patent with broad scope claims will have strong asset value potential.

There may be a few other reasons for filing the permanent application from the onset as well, but these are uncommon. For example, if interference is imminent, there may be value in filing promptly.

ILLUMINATED HAMMER

New Discoveries:
In our original design illustrated in Chapter 1, we started with a hammer with a light attached to the head . . . on the outside. We realized that this would simply not take the beating and could not come up with any viable alternatives to help cushion the light. This is where we made the discovery of having an internal light, which would be easy to cushion or may be made of fiber optics; by using modern plastics, we devised a hammerhead cap that is both durable and of high clarity material. If we had filed a patent application on the first version, we may have abandoned it based on the superior attributes of the subsequent discoveries.

If your patenting approach requires immediate filing of the permanent patent application, then the teachings of *The Patent Writer* become more important. Use it to write your patent applications promptly, as your patent attorney would most likely have to schedule the time to do it. You will also be in a position to maximize your patent protection by applying the tenets we've described.

PROVISIONAL PATENT APPLICATION STRATEGIES

There are several strategies you can employ when filing provisional patent applications. Surely you'll want to incorporate some of these in your plan.

Best Starting Point

There are many good reasons to make a provisional patent application your starting point, keeping in mind the big picture and your objectives. That is, to make money on your inventions—not just spin your wheels filing patent applications that may or may not have any commercial value.

From this perspective, product development is an essential component to writing successful patent applications and securing broad patent protection. You'll want to know how to effectively use this flexible, powerful tool to your advantage during the initial stages of product development.

Coordinating With Product Development

Unquestionably, the most desirable elements and attributes of an invention emerge during product development. Those elements and attributes that show the most promise from a marketing perspective are the ones you want to patent. We don't always know what those elements and attributes are in the first stages of development. They may not even emerge until much later on, when test marketing.

Follow *The Patent Writer's* tenets and you will become proficient at writing PPAs during the course of product development. You'll be able to change, add, and modify the product as it goes through its metamorphosis, and you'll be able to include all the newly discovered inventive subject matter—whether it is in a single provisional application or subsequent ones. Since PPAs are low cost and can be quickly prepared, you'll have

---

PATENTS = INVESTMENT MONEY

Many potential investors want to see a broad patent portfolio containing a number of patent applications and patents. Without a solid patent portfolio, investors may be leery about investing in a technology that may be easily copied by competitors.

COMMON MISTAKE

Many inventors feel that drafting a permanent application with claims is not significantly more difficult that drafting a provisional application. Besides the fact that preparing the claims of a permanent application can be very difficult, even for experienced inventors, you will also have to prosecute the permanent application with the USPTO, which usually requires the expertise of a qualified patent attorney.

much flexibility in declaring your patentable subject matter. Trying to do the same with a permanent application would be extremely costly and time consuming.

New Discoveries

Any important discoveries or new inventive matter that was not disclosed within a previously filed provisional application would be considered new subject matter. Thus, in order to establish a priority date on the new subject matter, you have two options. In either option you would not receive the benefit of the previous provisional application's filing date.

If you have filed a provisional application and new subject matter is discovered, here's what you can do:

PROVISIONAL DRAFTING TIP

While it is important to disclose as much as you can in your provisional application, you should be careful not to include subject matter for other inventions in the PPA. If you have multiple inventions, you should file multiple PPAs accordingly.

1. You can quickly and inexpensively file another provisional patent application on the new subject matter establishing the new priority date for the new discoveries; or

2. You may include the new subject matter when the permanent application is filed, which then establishes a new priority date for the new material.

We know that if you filed a provisional patent application prior to a public disclosure, you are protecting your worldwide filing rights. However, if any new subject matter has been publicly disclosed, and is being introduced into the permanent patent application, it can affect foreign filings. You should discuss this with your attorney, as a foreign filing that references the permanent patent application would be limited to only the prior (non-publicly disclosed) subject matter contained in the original provisional application, and not to the newly added material.

This rule regarding priority dates also applies to permanent applications. In other words, you cannot add additional content in a continuation application and necessarily preserve your filing date. Once again, discuss it with your patent attorney to decide on the best approach.

What To Include

Generally speaking, you want to include as much information as possible on a given provisional patent application so that when the permanent application is filed, you can determine what to include and what to delete.

You want to include as much information that is required to provide support for a later filed permanent patent application.

In other words, you'll usually want to draft your provisional patent applications as broad as possible. This is generally true regardless of whether the first patent application filed is a provisional or permanent application. Either way, it is an easy task to incorporate in all your patent applications.

In the *New Railhead Mfg.* case—*New Railhead Mfg. Co. v. Vermeer Mfg. Co. & Earth Tool Co.*, 298 F.3d 1290, 63 U.S.P.Q.2d 1843 (Fed. Cir. 2002) (download at www.patentwizard.com/htmls/newrailhead.pdf)—the patent applicant filed a weak provisional application and then later filed a permanent patent application that had new subject matter not contained within the provisional application. The resulting patent was determined *invalid* because the provisional application did not have sufficient disclosure to support the claimed invention in the resulting patent!

Keep in mind that it's recommended to have a patent attorney review even a PPA since there is a risk of having the patent office, or worse, a judge or jury, decide that your PPA does not adequately disclose the invention. In such a case, the validity of your permanent patent may be called into question.

When new subject matter is discovered after you've filed the permanent application, you then have the additional option of filing a new patent application covering only the new subject matter, which may be either a provisional or permanent application, or you may consider filing a second patent application as a "continuation-in-part" (CIP) of the first application. Writing and filing a CIP is discussed in Chapter 11. A CIP will have a priority date; however, the priority date will only apply to subject matter commonly disclosed in both patent applications.

Keep in mind that it's always easier and more cost effective to pare down a patent application than to add onto it.

ILLUMINATED HAMMER

New Discoveries: In our application we identified four inventions, the hammer itself, the method in which it is used, a process of manufacturing it, and a special composition used to form the hammerhead cap. You'll read more about how to identify this patentable subject matter in the next chapter.

Filing Multiple Applications

Another strategy you may employ by using provisional patent applications is to file several applications on the various inventive matters, instead of just an all-encompassing one. For example, let's say you have an invention that contains five different inventions all in one. It may be an apparatus that has three interrelated, innovative components, all of which are

A later-filed permanent application is entitled to the filing date of the provisional application only for "common subject matter" disclosed in both applications. In other words, if you add new subject matter to the permanent application that was not disclosed in the provisional application, this subject matter is not entitled to the filing date of the provisional application.

patentable, is made by a certain new process, and is used as a certain new system. Instead of filing one jumbo provisional patent application, you file five separate ones.

It may cost a bit more to file the five separate provisional patent applications, because the filing fees will total $500 instead of $100. But there are some advantages to doing this.

- **Marking "Patent Pending."** Marking products with a notice stating "5 Patents Pending" is a lot more powerful than just saying "Patent Pending." It sends a very important message to competitors.

- **Focusing.** Five separate provisional patent applications are easier to write. It really is easer to focus on a single invention and keep the subject matter organized. Trying to reveal several inventions in a single jumbo patent application can be cumbersome and confusing. The organization of multiple inventions is difficult to monitor and maintain. Filing separate provisional applications also keeps the inventions separate from one another for licensing and sale purposes (you may want to license inventions A and B but not inventions C, D, and E).

- **Tracking.** Separate provisional patent applications subsequently followed up with separate permanent applications are usually easier to track through the patent office. At first you'd think that filing five separate applications could end up being costly, but that may not be the case. Frequently, related patent applications will be given to the same examiner, or perhaps two different ones within an examiner's technology group. They tend to get reviewed in sequence, not all at once. The timeline from prosecution through issuance of the first application being reviewed is usually about the same or perhaps a little faster than if you file one jumbo permanent application, which is split out into divisional applications on the various inventive matters. This is discussed in Chapter 11.

Having several patents on an innovation may be an important strategy towards protecting it and licensing it, if that's your direction. With several patents pending, it makes licensing easier as you may elect to license them in one or more combinations and in one or more fields, perhaps some exclusively and some non-exclusively. Separate patent applications tend to provide greater flexibility to do so than one large one.

Keep the following in mind if you proceed with multiple PPAs: You still only have twelve months to file your permanent applications for each of the PPAs. Check with your attorney to determine his or her workload as well as the estimated costs of following through with the permanent applications. Five PPAs could translate to five permanent patent applications at a cost of $25,000 or more.

Review your patent strategy and budgets before you start the clock ticking on your provisional patent applications, and ensure that you have the budget and marketing/licensing strategies in place to begin deriving revenue from your patent investments.

WRITING A PROVISIONAL PATENT APPLICATION

The section of law that establishes the content required in a provisional patent application essentially defines a similar format for the permanent patent application with the exception of the claims, prior art disclosure, abstract, and the oath of inventorship. In other words, an application written as though it was the permanent patent application may be filed as a provisional application by leaving off these parts of a patent that are required for permanent applications.

Format

Since the format for both the provisional and the permanent patent application are the same, with the exception of the claims, the abstract, and oath of inventorship, a well-written provisional application may be converted into the permanent application fairly easily by adding the three additional parts. When you use the format in Chapter 8, however, you'll want to include a few additional considerations. They are:

1. **Use the same guidelines for line spacing, page type, margins, and so on as the permanent application.** This is a requirement, although it is frequently overlooked. If your provisional patent application doesn't "look like a real application," you are opening up the potential of it being scrutinized for other errors, and it may be rejected. Usually the Patent Office won't reject your application if there are some minor errors and omissions. However, it'll be readily received, filed, and a return receipt will be promptly sent back, if it "looks right."

2. **Write your provisional patent application as broad as possible.** You

TAKE PROVISIONALS SERIOUSLY

A provisional application requires the same level of effort and quality as a permanent application. Do not make the mistake of writing a poor provisional application—this can be very costly later when you are attempting to utilize the provisional's filing date in a later-filed permanent application for priority purposes.

may do this by including not only the patentable material you believe should be included, but other fringe considerations that may not necessarily be considered patentable. For example, a laser skin removal process used to extract cancer cells from a patient in a hospital may also include using the same grafting process for other applications. That may include its use for extracting a multitude of other skin abnormalities as well. It could also include using the process on animals, perhaps even plants. When filing the permanent application, you can either file it with the broadened scope, or there are several tactics you can employ to write additional patent applications covering other specified uses—for example, a CIP application for use with plant grafting. Even if your company is not going to use it in its field, it could be an excellent opportunity to garner additional profits by licensing it to other trades.

INVENTIONS ARE NEVER "DONE"

Every time you find an improvement to your basic invention, write it in your journal, and discuss these improvements with your attorney. Additional value may be added to your current patent application, or you may have the groundwork completed for your next patent.

3. **Drawings must sufficiently reveal the inventive matter.** A common error that is made with provisional applications is that the author may use inferior drawings that do not adequately depict the invention contained therein. If they don't clearly depict the inventive matter in the provisional application, it may invalidate the permanent one. With this said, you have two options. First, you may include professionally drafted drawings, such as you would with the permanent application, or secondly, take extra care in qualifying that your drawings do indeed clearly reflect the content. Generally speaking, most experienced engineers and inventors will prepare their own drawings—not have them professionally prepared—but will double check them to ensure they clearly illustrate the invention.

4. **Reveal alternative methodologies that may be anticipated even though they may be inferior.** By doing this, you are essentially "cutting the forest and leaving only your tree." The contents disclosed in the provisional application will become a public disclosure once your patent issues. You are, in essence, destroying others' ability to patent the alternate methodologies, as it will then be considered prior art.

5. **Include a provisional patent application cover sheet.** This is a simple form that lists the inventors, their addresses, the number of pages of specifications, and separately, the number of pages of drawings. It takes about five minutes to fill out and may be downloaded from the U.S. Patent Office website (www.uspto.gov).

6. **Include a filing fee.** For large entities, the USPTO filing fee is presently $200, for small entities (including inventors), the fee is only $100. If you forget to include the filing fee when filing the provisional application, you will still receive your filing date but you will have to pay a surcharge in addition to the filing fee. Visit www.patentwriter.com before filing each of your patent applications to ensure you are including the correct filing fee.

Granted, the U.S. Patent Office tends to be more forgiving on the format of provisional patent applications it has received, but why take a chance? It requires just a bit more effort to write a PPA that has all the scope and power you're pursuing, but without raising any red flags.

Should a "Claim" Be Included?

Many self-drafters have heard that they should include at least one broad claim within their provisional application. The reasoning is usually stated as adding a claim within the provisional application will reserve foreign patent rights.

However, a claim is not required by the USPTO for an applicant to receive a filing date or patent pending status with a provisional application. In addition, claims are difficult to prepare, thereby reducing the benefit (and purpose) of filing a provisional application in the first place.

It should also be noted that Section 4801(a) of the American Inventors Protection Act of 1999 amends 35 U.S.C. 111(b)(5) to provide that "notwithstanding the absence of a claim, upon timely request and as prescribed by the Director, a provisional application may be treated as an application filed under [35 U.S.C. 111(a)]." This clearly supports the position that a provisional application should be considered as a permanent patent application for foreign priority purposes.

Also, *World Patent Law & Practice* (Baxter, December 2001), a well-known authority on international patent law, states: "The World Intellectual Property Organization (WIPO), the responsible administrative agency for the Paris Convention, and the European Patent Office, moreover, have taken the position that provisional United States patent applications are regular national filings under Article 4 of the Paris Convention and are as a consequence adequate to establish a priority date."

There is also a potential downside with filing a provisional applica-

CLAIMS NOT REQUIRED

One of the significant advantages of provisional applications is that they do not require the complex claims section. By filing a provisional application without a claim, you can avoid having to learn the law on preparing solid claims. Also, do not just draft a quick claim for your provisional just for the sake of including one in it—this may cause more problems than it is worth.

"FILE WRAPPER ESTOPPEL"

You will hear the term "file wrapper estoppel" (a.k.a. "prosecution history estoppel" or "*Festo* type estoppel") used by your patent attorney to identify the limits of protection a patent offers. Any amendment or statement to the USPTO by a patent applicant that is material to the allowability of the patent can be used later to limit the scope of patent protection.

tion with one or more claims. If a provisional application containing a broad claim is redrafted as a permanent application (or a continuation application is filed), it is likely that the subsequent application will include different claims from that initially filed in the provisional application. A patent owner may be "estopped" (under *file wrapper estoppel*) from adopting a broad construction of a claim under the *Doctrine of Equivalents* (see glossary) if an infringer can prove that the subsequent claims amended the broad claim in the provisional application.

In view of the potential risk of file wrapper estoppel being applied to your future claims, and with no apparent benefit of including a claim, it is recommended that you should not include any claims within your provisional application. Of course, if you file a permanent application you will need to include one or more claims!

As a final word of caution: any patent attorney can tell you that a single ruling by a court can immediately change the patent law landscape—as has happened in the *Festo* and *Phillips v. AWH* cases (see Chapter 11). Thus, for the above statements regarding foreign patent office acceptance of the priority date of a PPA without claims, the law is relatively new, and has not been tested in the courts (nor has it been contested). In the event that a court ruling in a foreign country disallows a priority claim of a PPA that does not contain claims, your strategy of filing a PPA without claims may render pending applications invalid. Discuss these possibilities with your patent attorney before making assumptions about international patent law. If your patent attorney insists a claim should be included within the provisional application, then it may be prudent to make the claim narrow in scope to prevent the application of file wrapper estoppel. We do want you to understand that there are arguments both for and against this practice, and only you and your patent attorney can decide the best strategy for your particular objectives.

PROVISIONAL PATENT APPLICATION TACTICS

During product development you may discover business or legal issues that may affect your patent strategy—as we just mentioned. In addition, a competitor may begin the development of an infringing technology, or a new patent may be issued after your filing date claiming your invention. Conducting frequent patent searches while your patent application is pending may help identify newly-issued patents that may be cited as prior art by the Patent Office, resulting in your patent never issuing.

Fortunately, with a provisional application filed, you can quickly convert it to the permanent application, prosecute it, and get the patent issued. During this time period, which could be twelve to eighteen months, you would probably consider letting the other company know that you have a patent pending on the technology or product it is developing.

Alternatively, you can file a PPA that claims only the elements of your invention that do not infringe the patent recently issued to the competitor.

Depending upon the nature of your business, you may offer your technology for license before your patent issues. There are many advantages of doing this. The primary advantages are avoiding prolonged legal battles, and standardizing an industry. Whatever your position, make sure your attorney is informed and plays a part in your decision.

There is no better way to become expert than to work through the patent search and patent application process. You'll learn industry jargon and terminology, the competitive landscape, patenting activity in your field, identify potential licensees, and more.

At a minimum, writing your own provisional patent application is a relatively easy way to become knowledgeable of the patent process and of patents in the field of your invention. After all, you are ultimately the one who must decide on the patentable subject matter included in the application—and how much money you are willing to spend protecting it!

Chapter 5

Patentable Subject Matter

This is a real eye opener . . . and an essential step to writing a successful patent application.

When your invention is sufficiently developed and the inventive matter fairly well understood, you need to convert it into patentable subject matter. Doing so frequently uncovers a potentially broader scope for patent protection.

Frankly, this chapter is extremely important to your patent-drafting effort. It will show you several different types of patentable subject matter and will provide you with the tools to broaden the patent protection in your patent application. You can say it is like having two or more patents all wrapped up into one.

Most innovators tend to think in terms of a single concept, a single product they want to patent. In this chapter we'll expand your thinking so you will think in terms of multiple inventions within the one innovation. It's easy to do once you know the rules.

Once you understand patentable subject matter, you'll not only broaden the scope of your invention, but you may be able to get a scope so great that even existing competitive products will infringe your patent if they are subsequently made or used in the same unique manner you've claimed in your application.

For most engineers and inventors who are familiarizing themselves

ASK "WHAT" IS PATENTABLE

A common mistake for inventors is to approach a patent attorney and ask, "Is it patentable?" What the inventor should be asking is *what* is patentable. Though the patent attorney will not be able to give you a definite answer, he or she will have an opinion as to what subject matter, if any, appears to be patentable.

with the patenting process, this chapter will be an eye opener and one you'll use throughout your career. You'll want to make these concepts indelible in your mind.

THE FIRST STEP

Before you start writing your patent application, you need to identify the patentable subject matter. This may be difficult at first, but as you become more experienced with the patent-drafting process, you will eventually be able to identify all the potentially patentable subject matter for your inventions in a matter of minutes.

The primary purpose of a patent application is to fully disclose and claim all the patentable subject matter of your invention. Without first identifying what you can patent, you will not be able to prepare an effective patent application. You don't have to be a legal expert to accomplish this objective, just knowledgeable on a few U.S. patent laws.

At the end of this chapter is a checklist you may use to help identify your invention's patentable subject matter. After reading the following sections, make sure you complete this checklist. It establishes the subject matter you'll be writing about and revealing in your patent application.

TYPES OF PATENTABLE SUBJECT MATTER

There are various types of patentable subject matter available for utility patent protection. Familiarize yourself with each of these and determine which types apply to your invention. The more you list, the broader the potential scope of your patent application and the issued patent when granted.

Keep in mind that all inventions may be entitled to protection within more than one type of subject matter. In fact, most inventions usually are patentable in two or more areas, but inexperienced inventors commonly don't pursue them. For example, you may be thinking in terms of a patent on the structure of the product itself, but did you know that you may be able to protect the method in which it is used, the process of manufacturing it, and the actual composition of the raw material it's made of? By doing so, it's like quadrupling your patent protection.

The following categories are supported by statute and legal precedence producing millions of patents on the various forms of subject matter. However, many of the terms we use here are used loosely throughout the field of patenting. It's not so important whether the patentable matter is called a method of use, a system, or a process; what's important is that you understand the nature of the various types of patentable subject matter. By applying these concepts, you'll have taken a first major step towards writing a successful patent application, and a potentially valuable one at that.

Structures

Structure patents are the most common type of patent received by inventors and are granted for various subject matters, such as articles of manufacture, consumer goods, and machinery. Structure patents typically protect the physical structure of your invention. Sometimes they are referred to as product, apparatus, or device patents.

Examples of structure patents are Edison's light bulb, the original paper clip, and Zip-Loc® bags. They are usually fairly straightforward to understand. Since they are usually easy to describe, they are usually the easiest to write a patent application on.

When you literally describe the physical properties of an invention and its inventive matter in a patent application, you are defining a structure as the patentable subject matter. In doing so, you want to be thorough. You want to make sure you identify and describe all structural aspects that are considered inventive. At times there may be two, three, or more elements of a structure that may be patentable. Include them all in your patent application.

Later on, in Chapters 6 through 8, you'll learn some simple methodologies to combine related structural aspects together, instead of describing each one specifically. For example, if the structure of your invention requires gluing two pieces together, but permanently riveting them is even better, you can refer to this as "adjoining the two elements by a means of adjoinment, such as, but not limited to glue, rivets, adhesive tape, and so on."

Structural elements and *structures used in combination* may also be patented. For examples, let's say you invented a unique, inexpensive, one-way valve for plastic coffee bags. It uses a unique internal flapper to allow air to flow in only one direction. There are three structures you should reveal in the application:

ILLUMINATED HAMMER

In our patent application, we describe and claim the structure of an illuminated hammer. We also describe the use of replaceable or interchangeable cap elements; thus, we may claim them as well.

1. The flapper;

2. The flapper as a valve;

3. The valve used in combination with a plastic bag.

Patenting only one structure tends to limit the scope. For example, if a patented hair styling blower dries hair 10 percent faster because of a certain comb attachment that extends from the tip and fluffs up the hair, what is going to stop someone else from designing around it with some other form of "extended attachment"? Or, how can you stop someone from putting a comb-like attachment on an existing hair dryer, accomplishing essentially the same objective? We'll discuss solutions to this problem later on.

Methods

Method patents define a unique system or method in which a product (structure) is utilized. While patents cannot be obtained on a commonly known product, a patent may be obtained when it is used in a certain novel, unique method. A unique method of use typically consists of a combination of components used together that result in a system. Method patents are also commonly referred to as "method of use patents" and "systems patents." These can be powerful assets, and at times may even blanket the use of prior art in your unique system.

For example, using the previous hair blower example, you can patent the hair styling blower with the extended attachment, but you can also patent the system in combination, such as a hair blower with an extended attachment suitable for fluffing hair while warm air passes through it. By claiming the method of use, you're also blanketing any other extended attachment on any other blow dryer. Plus, you're also patenting the use of an extended attachment a competitor might sell to use in combination with an existing (prior art) blow dryer.

A method patent may also include the combining of ingredients or parts that, when assembled or used in combination, produce a certain outcome. A classic example would be the method of adding a catalyst to epoxy resin, thus causing it to harden. It may also include the application of component parts that, when used in a predetermined sequence, produces a desired result. For example, teeth may be whitened by following this

PATENTABLE MATTER

Be sure to discuss with your attorney different forms of patentable matter related to your invention. Very often, your utility patent may include not only the patentable features of your invention, but also the method of production, or other processes involved with the creation of the product.

sequence: 1) brushing the teeth with low-acid toothpaste; 2) rinsing with a neutralizing solution; and 3) applying the bleaching paste. In such a case, the individual components—the low-acid toothpaste, the neutralizing rinse, and the bleaching paste—may also be patentable if their ingredients are considered unique.

Most commonly, a single patent application is filed that describes both the structure and the method of use. This is usually the desired approach instead of splitting the two out into two applications, but you may want to discuss it with your legal counsel first.

Make sure you don't overlook this important facet of patenting—that is, incorporating methods of use in your application. Because if you do, and you launch or try to license your structure-only patented products later on, all you may be doing is showing a competitor or potential licensee how to design around it.

Business Methods

Business methods were not previously considered patentable in the United States, but all that has changed since the *State Street Bank v. Signature Financial Group, Inc.* decision in July 1998. Business methods patents are frequently related to Internet use means or computer related software programs, although this is not a prerequisite. Training programs may also fall under this category and are patentable. The Amazon "One-Click" patent is an example of a business method patent that also includes the accompanying software.

An example of a training-related business method patent might be something as simple as the use of a log book to record output so a trainee may compare his production results to a company's standard.

While there may not be much opportunity to include a business method patent in your application, if you're patenting something that is related to how a product is used in commerce, there may be an opportunity to expand your patent's scope. If it is software related, then you want to claim the software invention as well.

Processes

Process patents generally refer to manufacturing processes. Examples are the process used in Eli Whitney's cotton gin, which extracted cottonseeds, and a high volume gas-injected ice cream manufacturing process. A process patent may also include "the process of reading DNA in genes," which some may consider as a method patent instead.

Process patents can be powerful assets, as they may result in an important cost advantage just like the cotton produced from Whitney's cotton gin or the ice cream produced from the gas-injection process. Can you imagine if all cottonseeds were extracted by hand? Or if ice cream was made in churns? Their retail prices would be exorbitant! There's also no disagreement on the benefits of the process of reading DNA and its positive effect on medicine, law, and justice.

Before you think about whether or not your invention may include a patentable process, ask yourself if there was some structural aspect of your invention that had to be overcome in the manufacturing process in order to make it cost effective, or perhaps to make it work at all? If you've introduced a unique process as the solution, it may be patentable.

Sometimes a patentable process may appear to overlap into the business methods field. For example, applying unique process controls in a manufacturing environment in order to control quality and output may be a patentable concept. Or, a manufacturing-related business patent might involve the use of statistical process controls (SPC) to control two separate, but reliant, processes in a manufacturing operation such that the operator may then determine what manufacturing adjustments may be made. Then again, the adjustments may be made automatically. Are these process patents or a business method patent? Frankly, it doesn't matter as long as you've identified the unique patentable subject matter.

When filing a patent application for a process patent, you can also claim any products made from the process. So, if you've devised a unique manufacturing process for ice cream, one of your claims would be a claim for "any ice cream made by the process." This may become a valuable claim in a patent infringement issue because it not only gives you rights to sue for infringement of the process, it gives you the right to sue all subsequent marketers and users of the products it has produced. Don't overlook this opportunity.

Machines

When various components are assembled together to perform a specified function, it's usually referred to as a machine patent. Many, if not all, of the components may be prior art. What's important is that the outcome is

---

PROTECT METHODS & PROCESSES

Protecting the structure of a mechanical invention is important. However, the manufacturing process to make the invention and the method of utilizing the invention may be protectable. Manufacturing processes and methods of use are often times overlooked by inventors and patent attorneys.

unique. A machine may be small, such as a Cuisinart®, or it may be large, like a massive rock grinder at a quarry.

Accompanying a machine patent are invariably one or more process patents. There may be additional structure, product, or device patents that are part of the machinery as well. Again, it's not important what you want to call these categories, but you unquestionably want to make sure that you've covered all the inventive attributes associated with the machine. Just like with process patents, you can also claim individually any product that is made by the machinery.

In summary, a patent on machinery may include: 1) the machine; 2) its processes; 3) various unique devices (or components) that comprise the machine; 4) those devices' processes; and 5) any products produced on the machinery or by using the related processes. You may file one application covering all the inventive matter regardless of whether or not the patent examiner will split them out later into two or more patent applications (you'll read more about divisional applications later). The result is that you may end up with two or three patents with the potential power of five or six! A powerful position? You bet!

Software

This refers to the software itself and how it works. It also refers to Internet programs. Some examples of software patents are those associated with mouse applications, the one-touch screen, Amazon's "One-Click" patent, and the blinking cursor developed by an independent inventor. When patenting in this field, keep in mind that it is rapidly changing. Software patents tend to become obsolete quickly, unless they are very broad and yet specific in scope like the one-touch screen patents.

When patenting software, you are usually patenting a process, most often illustrated by flow charts. The object of software is to convert code into certain tangible results. For example, if you enter your annual income into tax calculation software, you may be able to print out completed tax return forms. The process between your entering your tax information and the tangible end result—the printed tax forms—could be the subject matter of a software patent.

If you're developing anything computer-related, your invention's patentable subject matter will tend to fall under this category (except for mask works, artwork for electrical circuits fixed in semiconductor chips).

However, make certain that you are including all the related patentable matter, such as: the software application; any applicable hardware (a device or machine?); any processes your software includes; and any products that may be an outcome of its use.

Chemical Compositions

Chemical compositions may be in a myriad of forms, such as the various types of plastics or genetic and biologically engineered substances. Chemical composition patents that are specifically related to the chemical composition itself are relatively difficult applications to prepare, compared to the other types of subject matter. If this is your field you are most likely a scientist specializing in a particular field.

However, there is another form of chemical composition patent application that you may consider. For example, when a special formula is used to manufacture a certain product in order to achieve a certain result, it could qualify as a unique chemical composition.

Take any number of products made from plastic that may require certain unique properties. For example, let's say that an invention requires a plastic with substantial rigidity for structural purposes, but with enough flexibility to prevent shattering if broken. If you have accomplished this by blending various types of plastic resins, you can patent that chemical composition—specific to the use in your invention and related products.

You should always look for these types of opportunities with your inventions. Surprisingly, many simple forms of chemical composition inventions are passed over by the inventor when all it takes is one easy extra step to add them to an application.

Some chemical patents may have a more complicated side, however. Some of these difficult issues are discussed in Chapter 11 under special considerations.

FIELDS OF INVENTION

Every invention falls under a "field of invention." The field of invention could be defined by various industry segments (dental devices, transportation), by certain technologies (such as silicon semiconductors, anti-rust coatings), by patent classifications, and so forth. In order to be successful with your invention, you must become an expert in your field of invention. Some of the major fields of invention may be obvious, but we'll cover a few of the more common ones here.

Electronics

Electronic inventions typically fall under structures, such as devices (a remote control), an apparatus (electronic wiring harness), systems (a DVD player's method of operation), and machines (a DVD player would be considered one). Electronics used in industry may also play a part in manufacturing process and machinery inventions. Certain chemical compositions may at times play an important part in a specialized electronic device's function. For example, a unique plastic formula with specialized additives that prevent microwave radiation.

Mechanical

Mechanical inventions will fall under structures (bicycles, tools, sewing devices, and so on), and very possibly may include an accompanying system or method of use. If the mechanical invention is used in industry, it may include an accompanying process application.

There is a broad array of other products that tend to fall under the mechanical heading as utilized by patent attorneys, even though they are not considered mechanical by nature. For instance, plastic grocery sacks, fasteners, non-computer hardware, even clothing, a soft sculpture, and so on. The term "mechanical" tends to be a catch-all for many inventions that are not otherwise listed in these fields.

Business Operations and Training

Business operations may include not only those related to the operations and training in a manufacturing environment, but to office administration methods, accounting, order processing, and more.

For example, certain accounting methodologies used in accordance with a proprietary, in-house, software application may very possibly constitute a business method patent. In such a case, there will most likely be an accompanying software patent that may be pursued.

Training applications may be very broadly interpreted to include the physical means of teaching others, conceptual and cognitive training methods, and even those methods used on the Internet.

Computers, Software, and the Internet

Of course computers would fall under a machine patent, but it goes on and on from there. Each individual component could also be within itself a device (like a hard drive), or apparatus (the mechanical system for inserting CDs), and may fall under any number of mechanical realms.

Computer systems by their very nature say that there is method patent potential. Regardless of whether the methods are created by use of software (like a computer program) or hardware components (like a keyboard and mouse), or even in combination, the resultant systematic usage can represent valuable patent potential.

Internet-related applications also have enormous potential. It is well-known that the Internet has created many millionaires—all of whom can credit their millions in large part to patent protection. Internet patents are predominantly business methods patents, unless they are accompanied by software and at times by hardware.

Biotechnology

This is almost exclusively the field of chemical compositions. However, there may be certain manufacturing processes that are used in order to create the desired chemical compositions that may also be patentable. Genetic sequences, tissue regeneration processes, and so forth are examples of biotech patents.

Often, pharmaceutical patents are found grouped together with biotechnology. However, as we well know, most drugs are actually chemical compounds, and therefore may more accurately reside in the chemical patents field.

Mathematical Formulae

Generally speaking, this is the field commonly using software applications that apply mathematical formulae to produce a desired outcome. It is quite possible, even probable, that a business methodology may result. Or, usage in industry may produce process patent potential.

Plant Patents

Plant patents are issued on any new plant species that is created asexually.

That means that the plants themselves do not create seeds that propagate the species. Examples of this include hybrid corn, tomatoes, strawberries, and more—but none of them will produce similar offspring.

Implied is the underlying suggestion that some process must be performed in order to create the species. This process itself may be patentable, and would fall under the definition of a process patent.

Anti-Terrorism Related

Whether this is a structure (a lightweight bulletproof vest), a method (the disarming of landmines), or a process (the x-raying of luggage at airports), the patent office will expedite your patent application with priority. All applications that deal with national security are subject to being put on the fast track of approval.

The United States Patent Office has a process called "make special." If you file a patent application for anti-terror related inventions with a make special request, your application will move to the top of the examiner's stack.

Keep in mind that the United States government has the authority to not issue or publish patents that affect national security. Therefore, you may never receive an issued patent; however, you may receive a reasonable royalty if the government elects to manufacture your invention.

Foods

This is an interesting and yet difficult field for patenting. Before pursuing a food patent, consider your options under trade secret protection. Coca-Cola® has been quite successful in building their product and brand around a trade secret—the formula for Coke® (see Chapter 1). Many, if not most, highly commercialized recipes and their related processes are maintained as tightly held trade secrets. But again, if it is going to be difficult to hold a trade secret confidential, then patent protection should be sought.

With foods, the ultimate intellectual property protection does not usually rest in patents or trade secrets. Trademarks have proven to be one of the stalwart forms of intellectual property protection for food products—as close to a century of marketing of Morton's Salt has shown. Patents may represent the best way to secure initial protection against knock-offs, but in the long run brand identification associated with a trademark is the IP of value.

RECIPES

Many recipes do not qualify for patent protection and should be protected as a trade secret. However, if your recipe has a functional aspect (reduces cooking time; prevents burning of food), then you will want to consider patent protection for your recipe.

81

Should you elect to secure patent protection on your foods, you should first consider the protection of the ingredients as a formula. It may be considered a chemical compound of sorts. Please keep in mind that a combination of ingredients that simply tastes better usually is not patentable—there needs to be some functionality (utility) for a food product to be patented. For example, does your new barbeque sauce reduce burning of the food? Better tasting alone usually is not grounds for patent protection. You need something more. That is why many recipes are protected under trade secrets.

Second, you should consider process patent protection in the unique way it is produced. For example, you may protect the ingredients of a salsa that has a very smooth texture and somewhat sweet taste—but also has a hot, spicy kick to it. You may also patent the process of how the chilies (for instance jalapenos and sereno chilies) are aged, thus making them very smooth tasting and perhaps even a bit sweet from fermentation.

Chemicals and Processes

The use of new, unique chemicals in industrial, commercial, or household applications may receive patent protection, regardless of whether it is a toxic chemical cleaner or a natural compound that repels bugs. Even a composition such as prior art powders blended to create a desired effect may be patented. When patenting chemical compounds, look for process patents as well. Frequently a process will be associated with chemical formulations, so you'll want to consider them too.

Chemical patents come into play in the fields of pharmacology, coatings, fragrances, food additives, rubber and tires, and so forth. Chapter 11 addresses some special considerations related to chemical patents.

Games and Entertainment

This may fall under two or more categories. First is the structure of the game's layout, the format. Next, you want to protect the method in which the game is played. Don't forget to add this important coverage because it'll make a design-around fairly easy by using a different format.

In addition, a game that you initially invented as a board game should include an accompanying software patent for the game's use as a computer game.

It is also important to secure copyright and trademark protection for board games and computer games. The copyrighted artwork and instructions should be protected—just look at Monopoly. A patent may serve as your initial protection, but like many other products, the long-term protection is your copyrights and trademarks.

Regardless of the field of invention in which you are creating, you will want to analyze the potential patentable subject matter based upon the checklist at the end of the chapter.

QUALIFY PATENTABLE SUBJECT MATTER

This is an important step. Once you have a basic understanding of patent law and you understand the various types of potential patentable subject matter that may be included in your application, you'll want to determine whether or not they qualify under U.S. law.

Once qualified, you'll be able to lay out a roadmap to writing your patent application. Here are three important qualifying points and an important definition you should know.

Qualifying Laws for Patentability

For an invention to receive utility patent protection, it must be novel, useful, and non-obvious. Any one patentable subject matter in your invention must satisfy all three of these factors.

Novelty

To receive utility patent protection, the invention must be considered novel (i.e. not previously known by others). This is a requirement of 35 U.S.C. §102. If the claimed subject matter within the utility patent application was publicly disclosed by another party prior to the inventor's provable invention date, the inventor will be barred from receiving a patent. Also, if the claimed subject matter was publicly disclosed by another party or the inventor more than one year before the patent application's filing date, the inventor will be permanently barred from receiving a patent.

A public disclosure by another party may be anywhere in the world at any time prior to the time you came up with the idea. This may include a patent, a published patent application, an article in a trade magazine, or of course, any product, method, process, and so on, that was produced and sold in commerce. It may even include a third party offering the product for sale.

Most inventions contain at least some subject matter that is novel. However, sometimes the novel subject matter is narrow in scope, which may not be in your best interests to pursue. A patentability search and a prior art search will assist you in determining whether your subject matter is novel or not.

Utility

MORE RESEARCH

At this qualifying point, you may be required to conduct more research, and of course, to conduct a patent search on the various types of patentable subjects you found that may be a part of your patent application.

The invention must also have some type of utility (35 U.S.C. §101). If an invention has a well-established utility, such as hammers, the invention automatically satisfies the utility requirement. However, if the invention does not have a well-established utility, for example, a new chemical composition, the applicant bears the burden of explaining to the USPTO that he or she has an invention that is useful.

For example, with a new plastic compound, the inventor must explain how and why the new composition has utilitarian value. Is it stronger? Does it process faster? Is it more flexible? What makes it unique and useful?

Non-Obviousness

Finally, the invention must be non-obvious to one skilled in the art of the invention at the time the invention was made (35 U.S.C. §103). The requirement of non-obviousness is usually the most difficult element to satisfy when attempting to receive solid patent coverage. If you are experienced in the field of your invention, you have an advantage in that you know many others who are skilled in the art.

One Skilled in the Art

"One skilled in the art" is any individual having ordinary skill in the area of technology related to the invention. A person of ordinary skill is presumed to be one who thinks along conventional lines of wisdom in the art, not one who undertakes innovation. In other words, the USPTO does not consider as being patentable any subject matter an expert in the field of the invention would deem obvious.

There are six factors considered in determining the level of ordinary skill in the art:

1. education level of the inventor;

2. types of problems encountered in the art;

3. prior art solutions to those problems;

4. rapidity with which innovations are made;

5. sophistication of the technology; and

6. education level of active workers in the field.

The subject matter sought to be patented must be sufficiently different from what has been previously used, known, or published so as to be non-obvious to a person having ordinary skill in the area of technology related to the invention.

For example, substitutions of one material for another or changes in size are ordinarily not patentable. However, if you can prove that using a certain color or reducing or increasing size has some unique advantage, which those skilled in the art have not previously considered, then it may be patentable.

For those familiar with ski racks that attach to car roofs, you are probably familiar with two companies: Thule and Yakima. Thule designed and patented a ski rack that held the skis at an angle of up to 29 degrees. It was said that Yakima, wanting to provide a competitive product, manufactured a rack that held skis at a 30-degree angle. Although there was little functional difference between 29 and 30 degrees, Thule missed preventing a competitive product by not patenting a larger angle.

PATENTABLE SUBJECT MATTER CHECKLIST

Now you should have a good understanding of the various forms of patentable subject matter. In order to maximize the potential scope of your patent application, it is time to analyze and qualify the various types of patentable subject matter surrounding your invention. Use the following checklist, take your time, and identify the various areas in which you may be able to secure patent protection.

WAS THAT OBVIOUS?

Every inventor (and patent attorney) struggles with what is *obvious*. Many inventors make the mistake of thinking their invention is obvious, and therefore not patentable, without even speaking to a patent attorney. What may seem obvious in normal terminology may not be obvious in legal terminology. Always seek the advice of a qualified patent attorney if you feel your invention is obvious.

Your Invention's Patentable Subject Matter

To identify your invention's patentable subject matter, determine which of the following applies to your invention by circling yes or no for each. A "yes" response indicates patentable subject matter.

Patentable Subject Matter	Yes	No
Structure: Identify what structural subject matter comprises your invention.		
A unique product, device, or apparatus?	(Y)	N
An element of the inventive structure?	(Y)	N
A combination of two or more existing products?	(Y)	N
Methods: Identify any methods or systems used in your invention.		
A system or method of use by an end user?	(Y)	N
A method of combining ingredients or materials?	(Y)	N
A system or methodology to apply components?	(Y)	N
Business Methods: Identify if there is a business-related application for your invention.		
A computer-related methodology?	Y	(N)
A method related to employee systems?	Y	(N)
A method of training?	Y	(N)
Processes: Identify any unique processes used for your invention.		
A manufacturing process?	Y	(N)
A process associated with machinery you've developed?	Y	(N)
Products made from the process?	Y	(N)
Software: Identify any software used by your invention.		
A computer program?	Y	(N)
Software used in a business method?	Y	(N)
Software used in a manufacturing process?	Y	(N)
Software used with a product, such as a game?	Y	(N)
Chemical Compositions: Identify if your invention uses any specialized chemical compositions or raw materials.		
A new compound with a specific use?	Y	(N)
A combination of existing compounds used in your invention?	Y	(N)
Total number of "Yes" responses/patentable subject matters: _____6_____		

Even if you are still uncertain of the patentable subject matter in your invention at this point, at least list the areas you believe may be patentable and let your patent attorney decide on the merits. Don't forget to research all the new types of potential subject matter you've encountered through a patent and prior art search.

For later reference, you should know that most of the items in this checklist would be considered independent claims in the patent application. Some may serve as dependent claims, for example chemical compositions. Right now, you do not need to be concerned about whether they are one or the other since you'll learn all about this in Chapter 6.

Once compiled, this list will be used to prepare a *claim statement* for your patent application. The claim statement will become the template for writing your patent application.

In this chapter, we have discussed what comprises a patentable invention. In some cases, your invention may be patentable in more than one type of subject matter, and you should take advantage of *all* that apply. Writing a patent that incorporates a structure, along with the method of manufacturing it, will result in a much more powerful patent. Be sure to constantly check yourself while writing your patent application, being sure that your invention satisfies the three patentability requirements: novelty, utility, and non-obviousness. Once you've mastered these elements, you'll be ready to start writing your all-important patent claims.

Claims—The Heart of a Patent

Stake your claim . . . and surround it with bulletproof protection.

The strength of your issued patent directly corresponds to the strength of your claims. Everything you write about in your patent application draft is geared towards bringing about one single outcome. That is, multiple, broad claims blanketing your invention from every possible angle.

It is recommended that you prepare proposed claims or at least identify the general subject matter of the claims you intend to have in your issued patent *before* you begin your patent application draft. Unfortunately, few inventors even consider claims before they write their patent applications.

It's obviously important to have broad, powerful, bulletproof claims to protect your invention and to protect against design-arounds. But how do you, a non-patent professional, accomplish this? Follow these guidelines and you'll see how you can develop a claim statement that will become the template for your patent application. It's not difficult to do, since you've just completed the *Patentable Subject Matter Checklist* in Chapter 5. However, before you begin, let's get a better understanding of claims in a patent.

OVERVIEW

Let's first review why we're trying to obtain a patent in the first place;

CLAIMS OBJECTIVE

We don't intend for you to become an expert writer of claims . . . that's your patent attorney's job. But it's important that you at least understand claim structure and you learn the terms that will broaden the scope of your patents.

it's to protect an invention. As we've previously discussed, the negative rights of a patent allow the inventor to prevent others from practicing the invention.

Although a patent contains many different sections, such as abstract, background of the invention, drawings, and so forth, it's actually the claims that identify the inventive matter that is ultimately protected by the patent. It's a requirement that every permanent patent contain at least one claim.

Therefore, once you have identified the subject matter that is potentially patentable for your invention, writing your claims will help you to precisely describe the inventive matter that you want to protect with your patent. With the patentable subject matter identified, it will then need to be articulated with carefully chosen words in one or more claims.

Many patent attorneys prefer to prepare the claims section of a patent application *first* so they have a solid grasp of what they need to fully disclose in the application. In other words, once an attorney fully understands the inventive matter you are trying to protect, he or she will then refer to the claims throughout the writing of the patent application. Once issued, the entire written portion of the patent will focus on reinforcing the claims.

As you read this chapter, you will see why your first draft of the claims is important for you to really understand the essence of the inventive matter that you are trying to protect. You'll see why it's critical to rely on your patent attorney to tighten your claims so that once your patent issues, it is powerful and defensible.

CLAIMS DEFINE PATENTABLE SUBJECT MATTER

A claim will describe subject matter that you are trying to protect. The novel and inventive elements of a patent must be described and contained in the patent in order to be enforceable. The descriptions of the inventive elements that are contained in the claims therefore describe the "patentable subject matter" of the invention. Thus, you can see the importance of establishing a claim statement to use as your template before you begin writing the description of your invention.

From a practical standpoint, if an invention becomes commercially valuable, others will try to find a way to trespass on the invention. That's called infringement. However, it's important to understand that infringement occurs only when the claims of the patent are infringed. We'll discuss later in this chapter what legal requirements must be met in order for an infringement to occur.

Every permanent patent application must contain at least one "independent" claim. The independent claim may be followed by one or more "dependent" claims associated with it.

Dependent claims place limitations on independent claims, thereby making them narrower in coverage than the independent claims they reference. All dependent claims should be grouped together with the claim or claims they refer to.

There is no limit to the number of independent or dependent claims a patent application may have; however, there is an extra fee for more than three independent claims and more than twenty total claims.

Each independent and dependent claim must specifically point out and distinctively claim the subject matter that will later be regarded as the invention. We've provided examples of independent and dependent claims later on in this chapter.

Taken one step further, it is paramount that a balance be struck between making a broad claim (which is sometimes easier for an infringer to argue against), and making too narrow a claim (reducing the scope of protection you desire).

In order for an infringement to occur, every element of a claim must be infringed. Therefore, a long, rambling claim comprised of multiple sentences would tend to be exceedingly weak, and therefore of little commercial value.

If a single element of the claim is not infringed, then the accused infringer should be able to defeat your charge of infringement. That is why an independent claim must be precise in describing the subject matter of your invention and must contain everything needed to properly describe the elements of your invention.

Avoid long, complicated-sounding claims, as they will probably contain subject matter elements that are not important to your invention, and should therefore be avoided.

As an example, let's fabricate a weak claim for the Illuminated Hammer invention. Let's say our Claim #1 was: "a hammer comprised of a striking surface, a battery, a wooden handle, and a light source that illuminated the item to be struck by the hammer."

Someone could conceivably *not* infringe if he or she copied every-

DON'T BE TRICKED

Even though a claim may contain several elements, don't be tricked into thinking it is automatically narrow in scope. Some patent attorneys will construct claims that may have eight elements, however six of them are part of prior art and only two describe the new inventive subject matter. In other words, there really are only two elements to infringe since all products in the field of the invention contain the other six!

thing described in our claim except that the person used a fiberglass handle. The inventor did not infringe *every* element of our claim. More on this later.

This is a simple example, but illustrates the importance of relying on well-written claims to obtain the best protection.

CLAIMS DEFINE SCOPE

The claims define the scope of the protection that you seek in your patent application. You will need to fully support the subject matter being claimed in the patent application through your written description of the invention and the drawings.

In other words, the claims may define an element related to a new automotive tire. If the patent fully describes vehicle handling in wet and rainy conditions, but fails to adequately describe how the specific novel elements of the tire perform, the claims would not be fully supported by the specifications in the body of the patent.

During the land grab of the old American West, the new landowners would "stake their claim" by pounding in marker stakes at the boundary lines of the property they were claiming to be theirs. Generally speaking, the farther apart the stakes were, the more land the owner could claim. Their land description would be "broader."

In much the same way, a patent claim identifies the invention owned by the inventor by describing the "metes and bounds" of the invention, or its intellectual property.

Broader claims that define broader scope can be significantly more valuable since they could describe more expansive rights to the invention. If you fail to fully support the claimed subject matter in your patent application, you may lose valuable patent rights.

The choice of wording of the claims can have a significant effect on whether a patent will be granted or not. Every word used in the claim could either narrow or broaden the scope of the patent, so when it comes to crafting powerful claims, be sure to rely on your patent attorney to write them for you. After all, this is where your patent attorney earns his or her money!

Nevertheless, it is *your* responsibility to prepare the scope of the claims in layperson terms by preparing a claim statement before writing your patent application draft. The claim statement is the template you'll

FOCUS ON THE CLAIMS

Claim structure and wording is more important in light of the *Phillips v. AWH* decision. As you proceed, be more concerned with identifying the potential claims you may obtain, than being able to write them technically correct. Let your patent attorney do that.

use to discuss and illustrate the patentable subject matter. Once completed, you may turn the claim statement over to your attorney to prepare the legal terminology on the permanent patent application.

Learn about the various types of claims well, and you'll soon be able to prepare your statement.

TYPES OF CLAIMS

As we've mentioned, a permanent patent must contain one independent claim, and may contain one or more dependent claims. The independent claims of a patent provide the "broadest" protection, while the dependent claims provide "narrower" protection.

When writing a patent application, it is advisable to include not only a primary claim that may succinctly describe the invention, but also a sufficient number of broader independent claims to completely cover your patentable subject matter. It is also important to include a sufficient number of narrower dependent claims as a safety net in case one or more of your broader independent claims should become invalidated, or if the patent examiner disallows certain claims to be included in the issued patent.

Dependent claims are also important during the prosecution of a patent application since their limitations can be incorporated with an independent claim to create allowable subject matter.

Independent Claims

A typical independent claim has a preamble followed by at least one element that defines the metes and bounds of the claim. An example of an independent claim is:

> *Claim 1. An illuminating hammer apparatus, comprising:*
> *a <u>handle</u>;*
> *a <u>head member</u> attached to said handle; and*
> *an <u>illuminating unit</u> connected to said head member.*

The underlined words are referred to as the "elements" of a claim, a total of three.

As we look at this claim, it's clearly minimalist in terms of the num-

UNNECESSARY ELEMENTS = PROBLEMS

After you write an independent claim, go through each element and check the elements that are absolutely required for a third party to manufacture and sell your invention. Put a check mark by the necessary elements and delete or modify the unnecessary elements.

ber of words used. That is, the claim contains everything that *must* be included to properly describe the invention, but contains no unnecessary words that could dangerously cause confusion as to the inventive elements that the inventors want to protect.

It would be difficult for another to make an illuminated hammer that did not contain a hammer handle, hammerhead, and something to generate light.

If we got carried away and wanted to impress our intelligence on the patent examiner or licensee (a trap that many inventors get caught up into) we might have added in unnecessary words to this claim, making it more difficult to enforce.

For instance, if we wrote this claim according to the following example, we might have very well opened the door for infringers to make and sell our illuminated hammer without recourse.

> *Claim 1. An illuminating hammer apparatus, comprising:*
>
> a *handle;*
>
> a ***claw hammer****head member* **glued** *to said handle; and*
>
> an *illuminating unit connected to said head member.*

With the added words "claw hammer" and "glued," we have made it possible for others to manufacture and sell an illuminated hammer with a *ball peen* hammer comprised of a ball peen hammerhead *wedged* onto the hammer handle. The unnecessary words "claw" and "glued" add limitations to the patent, narrowing the scope, and lessening the potential commercial value of this invention.

Remember, you want to be able to sue the manufacturer of an infringing product for literal infringement, not the reseller or end-user. Hence, it is important to craft your independent claims so that a product manufactured by a competitor would infringe upon your patent. For example, does your product require a battery to be included from the factory? Probably not. Do all of the structural elements require attachment to one another from the factory? Possibly not. If you want literal infringement of your patent, you need to identify only the absolutely required elements for your invention to be manufactured.

Dependent Claims

A typical dependent claim also has a preamble followed by at least one additional element. The dependent claims "depends" on a prior independent claim. Hence, the limitations of the prior independent claim are in effect, combined with the limitations of the dependent claim.

An example of a dependent claim is as follows:

> *Claim 2. The illuminating hammer apparatus of Claim 1, wherein said head member is glued to said handle.*

This Claim 2 is "dependent" upon Claim 1. The word "glued" is the additional limitation contained within dependent Claim 2. The scope of protection offered by Claim 2 is merely the combination of the elements within Claim 1 combined with Claim 2.

Stated another way, Claim 2 could be written as an independent claim, such as:

> *Claim 2. An illuminating hammer apparatus, comprising:*
>
> *a handle;*
>
> *a head member attached to said handle; and*
>
> *an illuminating unit connected to said head member, **wherein said head member is glued to said handle**.*

SAFETY NETS

Dependent claims are like safety nets to protect you if your independent claim is invalidated later. If an infringer is infringing upon an independent claim and a dependent claim, they need to invalidate *both* claims in order to avoid infringement. Using many dependent claims is a good strategy for reinforcing your independent claims.

YOUR CLAIM STRATEGY

The previous pages walk you through the mechanics of developing independent and dependent claims. However, these mechanics overlook an important component of a patent, or claims development: strategy.

A number of higher-level decisions go hand-in-hand with developing your claims, including:

- Cost considerations (you pay separately for claims that exceed three independent claims, and for more than twenty claims total)

- Claiming multiple inventions in a single patent (the foundation for splitting out the second invention later on in a divisional patent application or continuation-in-part "CIP")

- Splitting claims related to separate inventions into separate patent applications, or filing co-pending applications that claim variations on the inventive matter

- Filing many claims as a strategy to prevent the requirement to amend claims (See *Festo,* Chapter 11 on page 189)

There are many other patent strategies that could play into your decisions. Some strategies can speed up your patent issuance, other strategies can slow it down, others can leverage your priority date across a number of separate patents. These strategies are discussed in Chapter 11.

Your patent attorney may, of course, explain these strategies and others that could play an important role in the content and timing of your patent applications. Our objective in *The Patent Writer* is to focus on how to write successful, powerful patent applications, not to get into a major dissertation on patent strategy.

Only Claim Patentable Subject Matter

Once you have determined the patentable subject matter of your invention, it is then important to determine how to protect this patentable subject matter. You can attempt to protect your patentable subject matter with one patent application, or you may want to file multiple patent applications to protect it.

The important message here is to not try claiming elements that are not patentable, even though they are key to the functionality or intended purpose of your invention. As we've already discussed, by including unnecessary information, you run the risk of having the patent examiner disallow your claim(s) or patent, and even if approved by the examiner, there remains the risk of having a competitor more easily circumvent your claims.

Understand Patent Infringement

In order to prepare quality claims, it is important that you fully understand what is required for patent infringement to occur. It is not enough that you have a patent on your invention—you need a patent with claims that protect the essential portions of your invention. Patent infringement occurs when an infringing product "reads on" all of the elements of your independent claim.

Below is an example that illustrates the basic concept of patent infringement (X indicates an element is present). Keep in mind that Claim 1 consists of elements A, B, C, and D.

Patent Infringement				
	Claim 1	Product 1	Product 2	Product 3
Element A	X	X	X	X
Element B	X	X	X	X
Element C	X	X		X
Element D	X	X	X	X
Element E				X
Infringement?	N.A.	Yes	No	Yes

Product 1 is infringing because it reads on all of the elements (i.e. elements A, B, C, and D) of Claim 1. Product 2 is not infringing because it does not read on all of the elements of Claim 1 (i.e. Product 2 does not have element C as is required by Claim 1). Product 3 is infringing because it has all of the elements of Claim 1 (adding additional elements to a product, such as element E, does not avoid infringement).

Independent Claim Strategy

A good claim strategy is to have at least one independent claim on each patentable subject matter you listed in the checklist in the previous chapter. Ideally, it will provide the broadest possible protection if approved by the examiner.

You may include additional independent claims that are narrower than the broadest independent claim. These additional independent claims should focus on the most commercially viable version of your invention. In the event that the patent examiner disallows the broader claim, then you have a solid, commercially valuable back-up claim in the second independent claim.

If you have decided to file one patent application with multiple inventions and the examiner determines that they are separate subject matters, he or she will say so. This will create a divisional application in which the split-out subject matter will be reviewed in your choice of sequence, but still use the same, original patent filing date as the priority date. This is discussed in more detail in Chapter 11.

Dependent Claim Strategy

Your strategy should also include having dependent claims that focus on the patentable limitations, and should not include non-patentable subject matter (such as material types, dimensions, etc.) unless they are crucial to the invention. Dependent claims are your safety net for your independent claims and can be extremely valuable in situations where your broader independent claim is invalidated. Dependent claims also provide claimed subject matter that can be amended into your independent claims in order to place the independent claims in condition for allowance.

Using the Illuminated Hammer illustration, a set of dependent claims may look like this:

> *Claim 1. An illuminating hammer apparatus, comprising:*
>
> *a <u>handle</u>;*
>
> *a <u>head member</u> attached to said handle; and*
>
> *an <u>illuminating unit</u> connected to said head member.*

> *Claim 2. The illuminating hammer apparatus <u>of Claim 1</u>, wherein said head member is affixed to said handle by an adhesive means.*

> *Claim 3. The illuminating hammer apparatus <u>of Claim 1</u>, wherein said illumination device is contained within said head of said hammer.*

> *Claim 4. The illuminating hammer apparatus <u>of Claim 1</u>, wherein said illumination device is attached to an exterior surface of said head of said hammer.*

> *Claim 5. The illuminating hammer apparatus <u>of Claim 4</u>, wherein said illumination device is attached to said exterior surface of said head of said hammer by an adhesive.*

By using this approach, the claim(s) that another could infringe upon could be Claim 1 by itself, Claims 1 + 2, Claims 1 +3, Claims 1 + 4, and/or Claims 1 + 4 + 5.

As you can readily see, Claim 4 cannot be separated as an independent claim since it requires the superseding Claim 1. Thereafter, Claim 4 can be used only in conjunction with Claim 1 since it depends on it.

Choosing what dependent claims to include in the patent can be as important as selecting the independent claims in the first place. The number of dependent claims included in the patent application could significantly effect the filing costs, but if the combination of the independent and dependent claims are deemed commercially important, then it's a smart strategy to include as many as necessary to secure your ability to protect the broadest possible scope of your invention.

There are other strategies associated with the dependent claims. Once you have worked through the independent claims and have prepared your claim statement, your patent attorney will be able to help you quickly create a priority list of dependent claims that you may consider including in your patent application.

USING A PATENT ATTORNEY

Hopefully, you are beginning to get a clear picture that at least two people are required to create a strong, defensible patent: the inventor and the patent attorney.

You've already discovered why it's important for inventors to write their patent application, as well as the first draft of the claims, the claim statement:

- it causes the inventor to thoroughly think through all of the possible embodiments before selecting the ones that are most likely to be valuable;

- it allows the inventor to think through the various patent tactics and strategies, becoming more proficient in the patenting process; and

- it allows the inventor to save considerable investment that would otherwise be spent on having a patent attorney write the patent from scratch.

However, powerful patents are more often the result of the efforts of both contributors: the inventor and the patent attorney. Here are some

CUT THE FOREST

Here is an interesting alternative you may use in your patent strategy. In the body of the application, disclose as many methods, versions, and embodiments, but only identify the most valuable ones as your *preferred embodiments* and make sure they are protected by your claims. By doing so, you are patenting the best approach to your invention, while you keep others from patenting alternatives. You can't usually afford to patent everything, so in effect, you are keeping the "best tree" standing, but you're cutting the rest of the trees in the forest so there is nothing left for others to patent.

key reasons why a patent attorney is a critical part of your invention and patent-drafting strategy, and why it's vitally important to budget for patent legal fees in the course of your invention development:

- Patent claims require the use of language that is functionally precise, yet legally meaningful if challenged in the courts. Patent attorneys are skilled in crafting claims that satisfy both requirements.

- The impact of the *Festo* decision in regards to validity of claims amended during the prosecution of the patent requires the claims to be accurate in the original filing of the application. Patent attorneys are well versed in filing applications in light of the *Festo* case.

- In light of the *Phillips v. AWH* case, claim-drafting language must reflect a broad and accurate scope as would be understood by one skilled in the art of your invention. A smart patent attorney will know precisely what to do to ensure your patent application will not be unnecessarily narrowly construed.

- Prosecution of the patent claims (prosecution is the negotiation between the patent examiner and the applicant on the allowance of the claims during the patent pending period) requires knowledge of the legal and commercial nuances related to legal arguments. Patent attorneys are skilled in the art of patent prosecution, and are vital to getting the patent office to allow your most important claims with little or no amending.

- In the event that you find your patent as a subject of legal or court action, you will need a knowledgeable patent attorney to help you defend (or assert) your position. The patent attorney who helped draft your powerful claims will be the best advocate to help you launch your legal strategy.

Claims are the heart of the patent. Save money wherever you can, but when it comes to creating the key to your invention monopoly, it's time to bring on the best legal support for the job. A relationship with a patent attorney could last throughout the term of your patent, so look at your patent attorney as more of a business partner than a necessary, paid service.

IDENTIFYING YOUR CLAIMS

Determining what subject matter your claims should cover can be difficult

THE FUSS OVER FESTO

In the case *Festo Corp. v. Shoketsu Kinzoku,* the U.S. Supreme Court ruled that if the language of a claim of a patent is amended due to patentability issues, the inventor would lose the full benefit of the doctrine of equivalents for his or her patent. In other words, if the patent examiner decides that one of your claims is too broad and requires you to narrow it (amend your claim), your ability to rely on the courts to allow an equivalent interpretation of the words you use in your patent will be severely restricted.

and confusing. However, if you have done your patentability research, understand the legal requirements for patentability, and have prepared your *Patentable Subject Matter Checklist* in Chapter 5, you should have a good idea of what you'll be claiming.

Remember that the claims you prepare measure the invention for determining patentability both during examination and after issuance, when validity may be challenged. The claims also determine what constitutes infringement. It is therefore important to ensure the claims are sufficiently identified for your patent application before it is filed with the USPTO.

There are two important points to keep in mind when you prepare your claim statement. First, you don't want to limit the claims based on adding unnecessary words and elements. Second, you want to make sure that you've adequately described and discussed the patentable subject matter in the body of your application (the specification). By not describing the inventive elements in the body of your application, you can't include them in the claims, or may lose validity if later challenged.

As stated previously, a claim recites a number of elements (a.k.a. limitations). The claim will cover (read upon) only those products that contain all such elements. It is therefore important to not include any elements that a competing product would not require to be commercially viable.

Said another way, if someone manufactured a hammer that contained a hammerhead, a handle, and an illumination device in the head, it would infringe our sample Illuminated Hammer patent. Those three elements are the minimum requirements needed to make an illuminated hammer; therefore, a hammer that contained all three elements would read on our sample patent. It would infringe.

If we added the requirement that the hammerhead was attached to the handle using adhesive (a fourth element), and another company manufactured a hammerhead, a handle, and the illumination device, but joined the head and handle using a wedge, then only three out of four elements would read on our sample patent and would therefore NOT infringe. Again, we lose by adding too many words—by identifying additional elements that constrain scope, not improve it.

Now that you see the importance of writing claims that are not limiting in scope, you should first identify the claims which drive the subject matter you will disclose in the body of your application. But first, let's go over claim structure and cover a few examples of how the various types of claims on the inventive matter may be written. Then you'll be able to prepare your claim statement.

CLAIM STRUCTURE

Keep in mind that your attorney will be writing the legal claims and that learning the exact legal structure is not your objective. Your objective in this section is to learn how to write basic claims in layperson terms, so that you may describe your invention sufficiently and then be able to discuss the subject matter in your application.

Creating "Families"

Claim grouping is basically creating individual "families" of claims based upon a single independent claim (the "parent claim"). Once you have identified the independent claims you want to include within your patent application based on the checklist in Chapter 5, you will then identify dependent claims (the "children") that are dependent upon each of the independent claims.

Each family of claims should be directed to unique and non-obvious subject matter of your invention. If it helps, attach a name to each family of claims to help you organize them. In just a few minutes you'll be able to prepare your claim statement based on using this simple family approach.

Independent Claim Structure

Independent claims provide the broadest protection within the patent. Each independent claim has a claim number, a title, and a plurality of elements. Each of the elements must be "connected" to at least one other element (either physically connected or operable in relation to at least one of the other elements).

Independent claims have the following basic structure:

> *Claim [NUMBER]. A [INVENTION TITLE], comprising:*
> *[ELEMENT A];*
> *[ELEMENT B]; and*

[ELEMENT C].

Dependent Claim Structure

Dependent claims are dependent upon a prior independent claim or a prior dependent claim. Each dependent claim has a claim number, a title, and one or more elements.

Dependent claims have the following basic structure:

Claim [NUMBER]. The [INVENTION TITLE] of Claim [INDE-PENDENT CLAIM NUMBER], comprising:

[ELEMENT D];

[ELEMENT E]; and

[ELEMENT F].

Consecutive Numbering

Claims are numbered consecutively (1, 2, 3, 4, 5, etc.). You should not skip any numbers when numbering your claims. In addition, you should imagine each independent claim as a "wall" wherein later dependent claims cannot pass over to depend upon an earlier claim.

SAMPLE CLAIMS

Let's now take a look at how the basic claim structure is applied to the various forms of subject matter.

Structure Claims

Structure claims are the most commonly used claim. Structure claims focus upon the structure/connections of the invention. Typically, this will refer to the structure of a product, device, or apparatus as explained in the previous chapter.

Below is an example of a simple structure claim for a hand sewing needle with two pointed ends and a central eyelet (Claim 1 of U.S. Patent No. 6,189,747):

Claim 1. A hand sewing needle, comprising:

*an **elongated body** having a first end and a second end,*

> **CLAIM TIP**
>
> It is sometimes helpful to list your independent claims first to identify your families. After you have identified them, you can then build upon these respectively with dependent claims for reinforcement.

wherein said elongated body is straight having a single axis extending between said first end and said second end and wherein said elongated body is a solid structure and wherein said first end and said second end each are comprised of a pointed shape; and

*an **eyelet** positioned within said elongated body between said first end and said second end.*

Method or Process Claims

Method or process claims are also commonly utilized to protect a unique method or process. This claim structure may be used with either a manufacturing process or a method of use (system) application.

Below is a method claim for a poker-type game (Claim 1 of U.S. Patent No. 6,042,118):

Claim 1. A method of playing a poker-type game comprising:

a) *providing a deck of cards;*

b) *shuffling said deck of cards;*

c) *placing an initial bet by each player;*

d) *dealing only two player cards face down to each player;*

e) *dealing only three community cards, wherein one of said three community cards is face down with the other two face up;*

f) *giving each player a chance to view only the two player cards and only the two face up community cards, then giving the player the opportunity to double said initial bet;*

g) *showing all of said community cards, thereby providing a hand for each player comprising said three community cards and said two player cards; and*

h) *resolving each player's bet based upon a predetermined payout schedule.*

New Use Claims

"New use" claims are typically a method (or process) claim as stated above, and are used with new methods of use. For instance, DEET (N,N-diethyl-m-toluamide) was developed by the military as an insect repellent over forty years ago, and is still utilized within insect repellents today.

An independent inventor determined a new use for DEET, in that DEET can be utilized as an herbicide. Below is an example of a new use claim that was granted by the USPTO to the independent inventor (Claim 1 of U.S. Patent No. 5,948,732):

> *Claim 1. A method of utilizing an herbicide to control undesirable plant growth without damaging desirable plant growth, comprising the steps of:*
>
> > *a) providing an herbicide comprised of N,N-diethyl-m-toluamide; and*
> >
> > *b) applying said amount of N,N-diethyl-m-toluamide to undesirable plants.*

Obviously, DEET as a chemical composition cannot be patented, since it was created by the United States military more than forty years ago. However, the new use for DEET utilized as an herbicide is patentable.

Composition of Matter Claims

A composition of matter claim focuses on a chemical or material composition containing a plurality of elements. An example of a composition of matter claim would be:

> *Claim 1. A topical composition for treating a sunburn, comprising:*
>
> > *an amount of lotion; and*
> >
> > *an amount of aspirin.*

This kind of claim may also be used in a supporting role to an independent claim by delineating it as such.

HOW TO WRITE YOUR CLAIMS

No one is expecting you to be writing in legal terminology, only in straightforward layperson terms—language your patent attorney can make legally correct. There are a few things you can do to understand the terminology of the claim language in the field of your invention. You don't

WORD TIP

While extrinsic evidence (dictionaries) is still important, the intrinsic evidence (patent itself; statements to USPTO by patent applicant) is the most significant source for determining the meaning of claims. Hence, be very careful with the wording you use in your patent application.

need to take any classes; you have all the language you'll need to know at your fingertips.

Review Claims of Issued Patents

When you perform your patentability research, you should find at least five or more patents that relate to your technology. These five or more patents will be used as representative models that you can refer to when writing your claim statement.

Try to find patents where the patent attorney's name is listed on the patent and the patent appears to have been well written. The patent attorney's name will be on the front page of the patent next to the italicized field: *Attorney, Agent or Firm.*

Why study patents that contain an attorney's name? For the same reason that we're telling you to use your attorney for the claims writing and prosecution: the quality and precision of a patent document filed by a patent attorney will teach the proper way claims should be written.

As you read these next few pages, see if you can identify how the attorney used specific language to narrow or broaden the scope of the claims.

AVOID A COMMON MISTAKE

Many inventors make the mistake of including material types (e.g. metal, wood, plastic) within their claims. The material type of your product probably can be easily changed and is not essential; therefore, you should typically leave out the material type in your claims.

Avoid Being Too Narrow

To be effective, claims must not be too narrow. For example, if your claims do not cover all patentable embodiments of your invention, your claims are too narrow.

Although a hammerhead and handle are elements obviously required to comprise a complete hammer, focusing entirely on the hammerhead and illuminating device, yet neglecting to include the handle, will result in a patented hammer that does not include a handle—a device that will have little commercial value.

In addition, unnecessary limitations may be contained within your claims that are not needed to receive patent protection. For example, this independent claim has an unnecessary claim limitation:

> *Claim 1. An illuminating hammer apparatus, comprising:*
> *a <u>wood</u> handle;*
> *a head member attached to said handle; and*

an illuminating unit connected to said head member.

By adding the unnecessary limitation of a *wood* handle, the scope of the claim is limited only to infringing products that have a wood handle, not products that do not utilize wood (metal handles, composite handles, plastic handles, etc.).

One way to help avoid having a patent with narrow protection is to include a number of claims of varying scope to ensure proper protection of the patentable subject matter. Your patent attorney will know what approach is best to pursue.

Avoid Being Too Broad

To be effective, claims must also not be too "broad." Examples of claims that are too broad are claims that would cover the existing prior art or subject matter not adequately described in the specification.

For example, if the patent application for the Illuminated Hammer discussed, throughout the specification and claims, all of the features commonly found in all hammers, such as hammer handle and hammerhead, in addition to the features exclusively found in the Illuminated Hammer, then the example claim that follows would be considered too broad, and most likely would be disallowed by the patent examiner. The description and claims in your patent must pertain only to the exclusive patentable features of the present invention.

> **CLAIM CLARITY**
>
> If in doubt, define it. Claim interpretation is focused upon how a *person of ordinary skill in the art* of the invention would understand the claim terms.

Claim 1. An illuminated tool apparatus, comprising:

   a <u>handle;</u>

   a <u>device</u> attached to said handle; and

   an illuminating unit connected to said work piece.

This could easily describe a lighted shovel, a drill, a screwdriver, or even a baseball bat, if the business end of those tools were considered to be the "device." This claim would be considered too broad, since it is not adequately limited to the intended invention—the hammer.

By including a number of dependent claims within the patent application, you can reduce the potential risk of having your claim protection invalidated.

Now that you are armed with the knowledge of claim structure,

107

WRITING TIP

Focus on making a list of your key claims, and don't get distracted by trying to write your claims using "legalese." Patent-claims language is difficult, and is best left for your attorney— just make sure that you clearly communicate with him or her. If your attorney has a clear understanding of what you believe are the most important elements of your invention, he or she will craft those strong, broad claims that are possible— using the proper language.

claim scope and breadth, and more importantly, the impact properly written claims have on the strength and value of your patent, it's time to prepare your claim statement.

PREPARING A CLAIM STATEMENT

You are now ready to prepare a claim statement you can use as the template for your patent application. This is an important step. Once this is completed, you're ready to begin organizing and writing your application.

Your claim statement is nothing more than the patentable subject matter listed on your checklist on page 86, which has been converted into claims written in layperson terms. Each item on your list represents at least one claim that will protect your invention. They are typically the independent parent claims in the families of claims in which you will develop the dependent claims, or the children.

Since every patent application is different, we can't tell you specifically how to write and prepare your claim statement. So, we'll provide an example based on our Illuminated Hammer apparatus. In our invention, we will start by listing the patentable subject matter (from the checklist) and convert this into the four families they fall under. The patentable subject matter (and four families) includes:

1. The illuminated hammer apparatus itself (a product patent).

2. A process of manufacturing (a process patent).

3. A method of use (a method patent).

4. A specified plastic material (composition of matter patent).

Now we'll take these four and develop them into our claim statement, and include several dependent claims. Notice how our claim statement illustrates the four preceding patentable subject matters as independent claims 1, 7, 9, and 12, respectively (shown in bold type). Thus, our claim statement may look something like this:

> **Claim 1.** *An illuminating hammer apparatus, comprising:*
> *a handle;*
> *a head member attached to said handle; and*
> *an illuminating unit connected to said head member.*

108

Claim 2. The illuminating hammer apparatus of <u>Claim 1,</u> wherein said head member is affixed to said handle by an adhesive means.

Claim 3. The illuminating hammer apparatus of <u>Claim 1,</u> wherein said illumination device is contained within said head of said hammer.

Claim 4. The illuminating hammer apparatus of <u>Claim 1,</u> wherein said illumination device is attached to an exterior surface of said head of said hammer.

Claim 5. The illuminating hammer apparatus of <u>Claim 1,</u> wherein said illumination device is controlled by a switch.

Claim 6. The illuminating hammer apparatus of <u>Claim 1,</u> wherein said handle is comprised of a power source.

***Claim 7.** A process of manufacturing an illuminated hammer, comprising:*

    a) threading a plurality of wires into a passage within a handle;

    b) attaching said wires to an illuminating unit;

    c) attaching said wires to a switch;

    d) securing an engaging head to an end of said handle; and

    e) securing said illuminating unit to said engaging head.

Claim 8. The process of <u>Claim 7,</u> including testing said illuminating unit.

***Claim 9.** A method of using an illuminated hammer, comprising:*

    a) grasping an illuminated hammer;

    b) activating an illuminating unit within said illuminated hammer;

    c) concentrating the illuminating light upon an object; and

    d) striking said object.

CLAIM DRAFTING TIP

The words you use in your claims should be used in light of their "ordinary meaning" to one skilled in the art. Do not get artful in your wording—keep it simple and use words that have established meanings in the art. Use dictionaries, encyclopedias, and treatises if you want to double-check the "ordinary meaning" of a word.

Claim 10. The method of <u>Claim 9,</u> wherein said activating switch is automated.

Claim 11. The method of <u>Claim 9,</u> wherein said illuminating light is adjustable.

Claim 12. *An illuminating hammerhead cap, comprising:*

a hammerhead cap, wherein said hammerhead cap is comprised of a polycarbonate material with a durometer rating of approximately between 65 to 75.

Claim 13. The illuminating hammerhead cap of <u>Claim 12,</u> wherein said hammerhead cap is threaded for securing onto a hammerhead.

Claim 14. The illuminating hammerhead cap of <u>Claim 12,</u> wherein said polycarbonate material is comprised of a reinforced glass material.

As you can see, our claims are neatly bundled into four families: Claims 1 through 6 (i.e. the *structure*), Claims 7 through 8 (i.e. the *process of manufacture*), Claims 9 though 11 (i.e. the *method of use*), and Claims 12 through 14 (the *composition* of the product).

Keep in mind that an invention may include several other patentable components as well. If this is so, list them all in your claim statement. For example, with our patent application, you might have chosen to also include an independent claim on the two following inventions: the illuminating unit; and a plastic hammerhead cap.

However, we have elected not to include these two components as individual claims. An illuminating unit is not unique by itself, since it is used in flashlights all the time. If we included such a claim, it would be restricted to the use in an illuminated hammer—which we've accomplished in Claim 1 anyway.

Likewise, there are already hammers with plastic hammerhead caps, albeit they're not used for the purpose of allowing illumination to take place. We could have claimed the plastic cap individually, specifying a dual purpose role (threaded cap for securing to a hammer *and* for allowing light to shine through). However, other than for use with an illuminating hammer, what purpose could it serve? From this perspective, it then makes no sense to claim a plastic cap separately, since it's already covered in Claim 1. However, in Claim 12 we claim a hammerhead cap with a certain durometer hardness for its dual purpose use, distinguishing it from prior art plastic hammerhead caps, which we do not claim.

With your claim statement complete, you've only one more dimension of patent drafting to learn . . . the precise use of language. From this chapter, it should be clear to you why successful inventors leave the final claims strategy to a professional patent attorney—it's downright complicated to write strong claims that will properly define the limits of your invention, yet still be legally correct in the event that your patent ends up being the subject of a patent infringement suit. A patent with properly written claims has an overwhelmingly better chance of surviving, while poorly written claims may result in the patent being narrowed, or worse, invalidated.

As we'll see next, even the individual words you elect to use could mean the difference between a strong, defensible patent, and a weak patent written using words that can be interpreted differently by different people. How does "hingeably attached" differ from "attached using a hinge"? In Chapter 7, we'll take a look at how to carefully select each and every key word you may want to use while writing your patent.

Wording Basics

You don't have to be an English language major to write a patent application . . . you only have to be willing to learn a few concepts and methods.

Writing skills are important when preparing your own patent application. If you have a difficult time articulating your thoughts and inventions using written words, you will have to work a bit harder to write your patent application.

Because of the unique requirements to precisely describe an invention in words used by one skilled in the art, even skilled English teachers (otherwise thought to be experts in English vocabulary and grammar) may have some difficulty writing a patent application.

Understanding how to accurately utilize words to describe your invention is fundamental if you are to prepare your own patent application. You need to be able to describe the purpose, structure, functionality, and operation of your invention in terms that do not negatively limit the scope of patent protection.

You should attempt to utilize words with broad meanings to one skilled in the art of your invention. You should use dictionaries, encyclopedias, and treatises to confirm your understanding of the meaning of words you utilize in your patent application. You should also define specific words to illustrate that the word may have various meanings. Words that unnecessarily limit your invention should be avoided.

CAREFUL WORDING

With the *Phillips v. AWH* court case decided July 2005, extra care should be taken to ensure that your inventive subject matter is properly described throughout the body of the patent application so that one skilled in the art of your invention would understand exactly what you intend as your invention. We'll show you how to maximize this potential.

USING BROAD TERMS

Generally speaking, you want to describe your inventive matter with broad terminology. At times, a single word may be sufficient to do this. However, a combination of words may do it even better.

For example, do you want to use the word "groove" to describe the bending point in a plastic sheet, or do you want to use the word "hinge"? Hinge is the broader description, since it would include a groove, a weakened line upon which bending or folding will predictably occur, a ridged line, a perforated line, and so on.

Similarly, a group of words may serve the descriptive purpose better. For example, a "dimple" around a rivet hole may transfer stress; however, words such as "stress transfer region" or "stress transfer portion" would provide a broader meaning. You then would describe some of the various forms of stress transfer means that may be used (e.g. dimple, reinforced bead, knurled circular edge, heat-treated portion, etc.).

Here is another example relating to our Illuminated Hammer invention. Stating that the hammer handle is "glued" into the hammerhead or that the handle is "wedged" into the head is limiting. A broader description may state that the hammerhead is "affixed" to the handle. The word "affixed" simply indicates that the two parts are joined together to comprise a complete hammer.

But even the word "affixed," if not consistently used throughout the application, may project a different meaning if elsewhere the term "removably attached" is used to suggest that the hammer handle can be removed from the head and replaced with a similar handle. "Affixed" does not necessarily suggest that the handle can be replaced.

Yet another common error committed by inventors who write their own patents is one in which the inventor describes how he or she built his or her invention or prototype, and not how the construction should be broadly claimed.

A case in point relates to U.S. Patent No. 6,550,725. This patent contains only one independent claim. Based on what we've already covered up to this point, you will undoubtedly be able to point out many, many obvious errors in this claim. Figure 7.1 is an illustration of the product patented in the '725 patent.

However, we're using this patent claim to point out certain words that make this claim so narrow, it renders the patent nearly worthless. We've

BE OBJECTIVE

It is important to be *objective* when evaluating your writing skills. If your writing skills are not good, be honest and seek help so your patent application fully discloses your invention.

NO LIMITS

It is important to ensure that one skilled in the art of your invention understands that your invention is not limited to the embodiment(s) shown in the specification. Always utilize language such as "preferably," "approximately," and "for example" to help avoid limiting your invention to a specific embodiment.

made the words bold that we feel are too limiting, and put our comments or replacement words in *italics*. We believe that our suggested changes to this claim would have made this patent stronger.

Figure 7.1 Figure from U.S. Patent No. 6,550,725

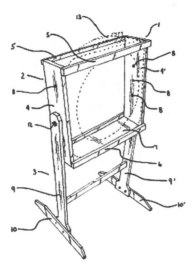

Claim 1. An adjustable fan stand for a standard box fan, comprising:

a) left and right corresponding **vertical** (*substantially vertical*) legs each having a **perpendicular** (*there is no need to include this limiting word*) foot support attached **at the bottom of** (*to*) the leg, wherein each leg has an upper fastening hole (*there are other ways to fasten*);

b) an upper fan housing, comprising left and right solid vertical sides connected at the **top** (*upper portion*) by upper front and back **horizontal** (*substantially horizontal*) braces and at the **bottom** (*lower portion*) by lower front and back horizontal braces and a lower horizontal bottom connected to the bottom of each vertical side; wherein, each of said vertical sides has a plurality of corresponding side adjusting holes located in the **center** (*leave this word out—"center" is a mathematical limitation*) of each vertical side at a plurality of vertical positions;

c) fan housing connecting means comprising a **wing nut** (*leave this word out since a non-wing nut fastener may easily circumvent this limitation*) fastening means, and

USING EXTRINSIC SOURCES FOR DESCRIBING TERMS

When writing the specification of your patent, consider identifying one or more specific dictionaries, technical dictionaries, and/or treatises that provide meanings for terms you want to define in your patent application. However, make sure to read all definitions provided by these extrinsic sources, and if any definition is inconsistent with the true scope of your invention, you need to consider either using a different term or clarifying the scope of that term in the specification.

115

    d) a lower, horizontal leg stabilizer connecting the lower parts of said legs together;

        whereby a standard box fan may be detachably inserted into said upper fan housing; and

        whereby said fan and fan housing may be adjusted vertically and tilted.

In short, you can readily see the words "vertical," "horizontal," "center," "top," and "bottom" are used to describe absolute positions. They are dangerously limiting, since slightly angled parts can easily get around the absolute requirement that the components be vertical or horizontal. "Substantially vertical" indicates that the member should be extending up and down, but there is not the necessity for the member to be absolutely vertical.

Also, by placing a hole one inch off "center," one can easily circumvent the limitation of the claim. Finally, the use of the word "wing nut" puts a ridiculous limitation in the claim since the intention of the claim is to allow adjustment—which can be accomplished through the use of a clamp, lever, spring actuated buttons, or even a nut and bolt (without the "wing" limitation placed on the nut).

By referring to these illustrations, you can understand the importance of selecting words that will project the broadest interpretation of the claims. Apply this approach to your first draft of your own patent claims and see if you can select words that accurately describe the important elements of your invention, without unnecessarily limiting the scope.

WORDS WITH ORDINARY MEANING

DON'T INVENT WORDS

Do not try to be your own lexicographer unless you simply cannot find a term commonly used by one of ordinary skill in the art of your invention. It is rare when an inventor cannot find a term commonly used in the industry to properly describe his or her invention.

In view of the *Phillips v. AWH Corporation* case decided July 2005 and prior case law, it is strongly recommended to utilize words in your patent application that already have an established "ordinary meaning" in the industry. Utilizing words with established meanings without redefining the words is referred to in *The Patent Writer* as the *ordinary meaning technique*.

The ordinary meaning technique is relatively basic and simple as suggested by its title. The patent drafter utilizes words that have established meanings to one skilled in the art of his or her invention. In other words, no conflicting meanings for words are used in the patent application that would give a word a meaning different than the "ordinary meaning." In addition to your written patent, technical dictionaries, encyclopedias, trea-

116

tises, and other types of extrinsic evidence may be used by a court to help determine the ordinary meaning of the words used in your application. Throughout the book we'll provide you with examples that will make it easy for you to follow.

The potential upside to this method is that it's simple, easy, and reduces the chances of including an unnecessary limitation within your patent application. The potential downside to this method is that you are stuck with the ordinary meanings of terms as understood by a hypothetical person of ordinary skill in the art of your invention.

Keep in mind that the ordinary meaning may be "trumped" by a clearly different meaning in the specification or by statements made to the USPTO during prosecution of your patent application. Attempting to redefine words (lexicography) in order to describe your invention has potential pitfalls for self-drafters. However, if done properly, being your own "lexicographer" can broaden or tailor the meaning of words in the claims for increased patent protection.

LEXICOGRAPHY

Lexicography is defined in *Webster's Revised Unabridged Dictionary* as "The art, process, or occupation of making a lexicon or dictionary; the principles which are applied to making dictionaries." Put another way, lexicography is the art of making up words, and applying a certain definition to those words.

It is well settled that a patentee can be his or her own lexicographer. In other words, an inventor may choose to create his or her own terms and use them as desired so long as they remain *consistent* in their use and their meaning is reasonably clear throughout a patent. Inventors are encouraged to first utilize only words that have an established ordinary meaning to one skilled in the art, and then determine whether they should attempt to modify this meaning through lexicography.

Also, an inventor may define his or her own terms, regardless of common or technical meaning and whether those words actually appear in any dictionary. While inventors may be their own lexicographers, a term in a claim may not be given a meaning repugnant to the usual meaning of that term (e.g. the word "hammer" cannot be used to define a "screwdriver").

According to *Lear Siegler, Inc. v. Aeroquip Corp.*, granting inventors the privilege of lexicographic license, that is, the right to define terms at variance with conventional definitions and usages, is done in order to hold

SAY WHAT YOU MEAN

It is important to utilize terms consistently between the claims and the specification. Any difference in usage may be utilized by a court to limit the scope of your claims.

117

open the possibility of obtaining a patent where an inventor is not schooled in the terminology of the technical art to which his or her invention pertains, or where there is a need to coin new expressions with which to communicate the invention.

One example that illustrates the benefits and dangers of creating new words is the phrase "hingeably attached." This phrase conjures up the image of one part that is attached to a second part by the use of a hinge—such as a door being "hingeably attached" to the door frame.

However, does "hingeably attached" actually refer to two parts that are attached, but which act like a hinge? Whatever connotation you attach to this phrase, you would need to first describe it clearly, then use the terminology throughout your patent application so that it consistently projects your intended meaning without confusion on the part of the reader.

The lexicographic privilege decreases clarity and certainty in patent claim interpretation. This occurs because it imposes on anyone evaluating a patent a burden of determining (through study of a patent's specification and prosecution history) whether the patent gives an unconventional meaning to a term that otherwise has an established meaning.

However, the burden would exist to an extent even if inventors were required to use conventional terminology, because the meaning of such terminology can change over time. One studying, in 1999, a patent applied for in 1980 and issued in 1983 cannot safely assume that terms in the patent have their 1999 meanings.

As an example, it would be unreasonable to expect to find words that describe an "online auction on the Internet" in a patent that issued in 1985—years before the Internet became a household term, and more than a decade before E-Bay popularized the online auction.

By reviewing the terminology used in recently granted patents, you will quickly increase the quality of your patent-drafting skills. Keep in mind that since words used in older patents may now have different meanings, or certain new words used in industry may have more precise meanings, it's important to employ those words that represent the most modern, applicable, meanings.

Therefore, as long as terms or words are used consistently throughout the patent application, and as long as those words project a clear and understandable meaning to the reader, they are allowed to be used—despite the existence of those words prior to the writing of the patent application.

PATENT DRAFTING TIP

Many inventors make the mistake of using patents granted more than ten years ago as a model. It is strongly recommended to use patents granted only in the last three years if possible because of changes in the laws and in the ordinary meaning to one skilled in the art. In addition, inventors should not use patents just because they are the longest. The length of a patent does not equal quality in patent applications.

INDUSTRY TERMINOLOGY

Many inventors do not learn the terminology utilized in the industry of their invention. It is important for you to research the terminology utilized within dictionaries, encyclopedias, treatises, prior art patents, industry publications, catalogs, and websites. Studying patents in your field of invention will accelerate your understanding of various methods of describing the inventive elements.

However, even within industries, different words can be used to describe certain elements. For example, United States patents for automotive devices filed by British inventors may use words such as "screen" rather than "windshield," or "bonnet" rather than the word "hood" to describe the part of the car body that covers the engine. Even "engine" may be described throughout another patent as a "motor," although the word "motor" is used in the electrical industry to describe an electrical device that contains a stator and rotor.

Read and learn the terminology used most consistently throughout your industry sector, and use those words consistently throughout your patent.

WORD TIP

By utilizing words consistently that already have an established meaning in the field of your invention, you will be assured that if your patent ever has to be interpreted by a court of law that it will *provide the meaning you intended*—and not a meaning you did not intend or were unaware of!

USING A THESAURUS

A thesaurus can be another valuable tool in determining the proper words and phrases you should utilize in your patent application. For example, you may want to initially utilize the word "bolt," but by using a thesaurus, you will find that there are broader words such as "fastener." Of course, make sure that the substitute word you find has an established broad meaning to one skilled in the art of your invention.

Many word-processing programs have a built-in thesaurus you can use (though it is limited). You can also utilize Internet-based thesauruses.

The thesaurus can also come in handy when first conducting your preliminary patent search. PatentCafe's ICO Global Patent Search (www.patentcafe.com) automatically generates a set of related thesaurus words for each patent that appears in the search results. Following a patent search for "running shoe," the Dynamic Search Thesaurus (DST) will show you related words of possible interest such as metatarsal, insole, over sole, instep, cleats, and so forth.

Use thesaurus words to expand your vocabulary in an effort to broaden the description and claims of your own patent.

ACRONYMS

You can greatly simplify your patent application by assigning initials to word groups that describe various parts of your invention. It is sometimes easier to write and understand a patent application that efficiently utilizes acronyms.

You can utilize acronyms that are widely recognized in your industry (e.g. HVAC—Heating, Ventilating, and Air Conditioning). You can also create your own acronyms (e.g. RCFM—Remote Controlled Flying Machine).

The number and types of acronyms you utilize is unlimited. However, you need to identify the particular acronym you plan to utilize within the application so there is no confusion, as some acronyms have more than one meaning. Be careful that you do not create an acronym that is widely used in an industry to describe something entirely different from your own intended use. See the Resource List on page 225 for more information.

In summary, while the structure of patent claims is critical to establishing the scope and breadth of your issued patent, the choice of individual words can have just as strong an effect on the strength or weakness of your patent. Remember also that just because you are allowed to "invent" new words, or your own lexicon, to describe your invention, venture cautiously. As you begin writing you application in the next chapter, be conservative, and err on the side of clarity.

Writing the Application

Get out the paper and pencil . . . start up the computer . . . and let's put your invention into words and drawings.

Writing a patent application can be intimidating. Even experienced patent writers have a sense of anticipation when writing about new subject matter. However, by following the structure and format of the application in this chapter, it'll become much easier to tackle, and in a short time you will have prepared a well-written patent application fully disclosing your invention. With experience, you'll be able to prepare patent applications and the accompanying sketches in a matter of hours!

To simplify matters, whether you are writing a provisional or permanent patent application, we know the format and the content in your draft are essentially the same. The only real decision you have to make is how it will be filed if it's a provisional application, or if you want to forward it to your patent attorney to have it completed and filed as the permanent patent application.

OVERVIEW

Patent drafting is like storytelling. It's a brief history lesson about the background of your invention, what's been used in the past, what's being used now, and last, how your invention overcomes problems with present-day products. As you write your patent application, you will be giving this history lesson and describing your invention in, more or less, a sequential timeline.

The Patent Writer format is simple to use and dovetails this brief history lesson right into your patent application, as it relates to the patentable subject matter and claim statement you've compiled. In doing so, it will be put into a proper, legal format with the right structure and language, and will be easy for others to read and follow. Then you will complete the patent application with detailed descriptions and drawings that properly describe the subject matter.

If you are using a tool, such as *PatentWizard* software (see Resource List on page 225), the method in this book will be of great assistance to you when preparing your patent applications. With the knowledge you've gained from *The Patent Writer,* you're armed with the ability to maximize your provisional patent application's strength and scope and quickly put it in a legal format.

Once you've completed a patent application written *The Patent Writer* way, it'll be a simple process to send it to your patent attorney for review, editing, and filing as a provisional patent application. Or, you may submit it to your patent attorney, who will then prepare the legal claims and file it as the permanent patent application.

STRUCTURE OF GRANTED PATENTS MAY DIFFER FROM PATENT APPLICATIONS

Remember, the structure you see in granted patents is *not* the structure utilized for preparing and filing patent applications. For example, the abstract is at the end of a patent application, while on the front page of a granted patent. Follow our instructions or consult with your patent attorney if you have questions.

ANATOMY OF A PATENT

Before you begin writing your patent application, you'll need to know the parts that comprise an issued patent. What you include in your permanent patent application will ultimately become these parts once filed, prosecuted, and granted.

Today's patent-drafting format is based upon laws and rules enacted in 1996. The format is a simplified version of the prior patent-drafting format, although the older one still covers essentially the same content as that required in the new format. The anatomy of a patent we describe herein is based upon the most recent laws. The sequence is that of an already issued patent, and will vary somewhat from the sequence you'll use to write your application. Usually, the abstract is written last, not first, and the drawings are prepared while you are writing your detailed descriptions.

Nevertheless, after you've read this section, you'll better understand the structure of a patent and how your patent-drafting format fits in with the content.

Table 8.1, Anatomy of a Patent, contains the various elements of an issued patent as it applies to patent drafting. This summary is based upon a patent on a relatively simple invention using the new format. This table represents the components for virtually all patents, whether they are products, processes, methods of use, software, machinery, and so on, or some combination of these.

Table 8.1 Anatomy of a Patent	
Page	**Elements**
Cover page	1. Title of the invention (a descriptive representation of the invention) 2. References (a list of other related patents cited and other prior art references) 3. Abstract of the disclosure (one page summary of the content) 4. A representative drawing of the invention (this is one of the formal drawings in the patent)
Drawing pages (this could be any number, but usually about 3–5 pages)	1. These are the formal drawings you will submit. They'll be illustrated in the sequence as you've submitted them. It may not include certain drawings that might have been withdrawn due to a divisional application or due to abandoned material.

Written pages (this could be any number, but usually about 10–20 pages). The subtitles in this part include:	1. Cross-Reference to Related Applications (this refers to any patent applications you may have previously submitted, such as a provisional patent application or any other co-pending applications, provisional or permanent)
	2. Background of the Invention
	    a. *Field of the Invention*
	    b. *Description of the Related Art*
	3. Brief Summary of the Invention
	4. Brief Description of the Drawings
	5. Detailed Description of the Invention
	6. Claims (the legal, binding claims approved by the U.S. Patent Office)

Provisional Patent Application Requirements

NEW DISCOVERIES

Remember, a big advantage to writing and filing provisional patent applications is that during development, should you make new discoveries, you can write and file additional provisional patent applications. When doing this, you can usually start out with the same computer draft from a previous application and change it accordingly. The newly drafted patent application can then be either incorporated into the final, permanent application or can be filed as a separate application later. Discuss this strategy with your patent counsel.

Table 8.1, Anatomy of a Patent, lists the components of an issued patent based upon the information provided in the permanent patent application. However, many, if not most, inventors and companies begin with provisional applications, which have a greatly simplified structure.

The U.S. Patent Office requirements for a provisional patent application, and acceptance thereof, include the following components:

1. *Written specification* adequately describing the invention and the preferred embodiments. This is defined in U.S.C. 112, first paragraph. This is what you'll learn in this chapter.

2. *Drawings* per 37 CFR §§ 1.81 and 1.83. Use the drawing format as defined here.

3. A provisional patent application *cover sheet.* You may download one from the U.S. Patent Office or our websites (see Resource List on page 225).

4. A *check* for the filing fee (presently $100 for small entities and $200 for large entities).

A Verified Statement claiming small entity status is no longer required. If you are a small entity, you only need to include a statement in the cover sheet that you are a small entity and pay the lesser of the two fees.

You'll soon find that writing a provisional patent application is a relatively simple procedure.

Patent Application Layout

As you know, writing a patent draft that will be filed as a provisional application or the permanent application uses essentially the same layout and format, with the exception of the claims section. Thus, your patent-drafting approach is greatly simplified.

Writing a provisional or permanent patent application draft is somewhat like writing a high school or college term paper, although it's most likely more technical in nature. Both types of patent applications should follow the same Patent Office rules with the correct layout, format, and content. A poorly written application with inadequate references and incorrect format may raise a red flag when received at the USPTO and get rejected. The following information lists the layout guidelines you should use.

1. **Overall Format.** Draft the written portion according to the standard manuscript format on $8^1/_2$" x 11" white, bond paper. Your lines should be double-spaced (although 1.5 spaces is okay and is actually easier to read), with 1-inch margins at the top, bottom, and both sides. Use an easily readable font, such as Times New Roman 12 point or Arial 11 point.

2. **Number Paragraphs.** Permanent patent applications (and provisional ones) should have all paragraphs numbered consecutively (i.e. [0001], [0002] . . .). One alternative to paragraph numbering is to have line numbers within each page (line numbers restart at 1 for each new page). This allows for quick and easy reference during the prosecution of the patent application. Indent the paragraphs.

3. **Headings.** The subtitle headings should be centered, in bold type for easy identification. When writing your titles, subtitles, and copy, use capital and lower case letters. Do NOT use all caps.

Drawings should meet the Patent Office's standards for number and letter references (this is illustrated in Chapter 9). If the drawings do not satisfy the USPTO standards, the patent application will not necessarily be returned to you, but you will be required to submit substitute drawings—so try to submit the correct drawings the first time.

Since it is important the permanent patent application reads on your provisional application, using the same layout and format also makes it easier for your patent attorney to convert later on.

Before You Start

In Chapter 5, you prepared a *Patentable Subject Matter Checklist.* In Chapter 6, you identified the claims you will be pursuing based on this checklist and prepared a claim statement. Your claim statement is your template for the subject matter you will be disclosing in your patent application.

If you have not prepared a claim statement, do not proceed until it's completed. You will be, in essence, cheating yourself out of maximizing the patent's scope and frankly, you'll be writing aimlessly. Is "ready, fire, aim" the approach you want?

IMPORTANCE OF PROPER STRUCTURE

Whether you are filing a provisional application or a permanent application, it is very important to utilize a proper and clean structure for your patent application. You want your patent application to look like it was prepared by a professional patent attorney and not by an individual who just threw some words together just to get "patent pending." Also, the overall appearance and structure of your application will create a "first impression" on the USPTO examiner, which is very important.

USE A COMPUTER

Without question, you should write your patent applications on a computer, using a word-processing program such as Microsoft Word. It is a simple task to prepare your layout, number lines, and so on. It also makes it easy to edit and manipulate text. You can even use the spell check and thesaurus option to help you find more descriptive, accurate words to use.

THE STANDARD FORMAT

The legal requirements you'll need to submit to the U.S. Patent Office are actually easier to write than it may appear. With your claim statement completed, follow these instructions carefully and you'll complete your application sooner than you think.

In 1996, the USPTO amended Rule 77(a), which provides that the elements of the *permanent patent application* should appear in the following order:

1. Title of Invention
2. Cross-Reference to Related Applications
3. Statement Regarding Federally Sponsored Research or Development
4. Background of the Invention
   a. *Field of the Invention*
   b. *Description of the Related Art*
5. Brief Summary of the Invention
6. Brief Description of the Drawings
7. Detailed Description of the Invention
8. Claims
9. Abstract of the Disclosure
10. Drawings

Likewise, a *provisional patent application* should have the following elements written in the following order:

1. Title of Invention
2. Background of the Invention
   a. *Field of the Invention*
   b. *Description of the Related Art*
3. Brief Summary of the Invention
4. Brief Description of the Drawings
5. Detailed Description of the Invention
6. Abstract of the Disclosure
7. Drawings

As you proceed from this point, write your patent application draft based on the provisional patent application format. The only decision you'll have to make after completing this draft is whether it is submitted as a provisional patent application or whether you turn it over to your attorney to have him or her complete it with the legal claims as the permanent patent application.

As you write your patent application, follow the patent application example of the "Lighted Hammer Apparatus." See pages 128–137 for the full sample provisional patent application of our Illuminated Hammer. Fill in the content of your patent application based on the instructions in the "You Write" tips located in the margins.

YOU WRITE

Use the instructions in these text boxes to write your patent application.

TITLE OF INVENTION

The title of your invention needs to *describe* what your invention is. The title should sufficiently portray the invention so that an individual reading it will be able to immediately identify what subject matter your invention relates to.

Descriptive Titles

For example, with our Illuminated Hammer invention, the title of the invention could be "Illuminated Hammer Apparatus" or "Lighted Hammer Device." We may also use a more generic term such as "Illuminated Hammer," so long as it is descriptive of the invention.

Similarly, if your invention is a process or method, you want to describe what the process or method is. For example, "Cotton Seed Extraction Process" (the cotton gin) or "Method of Reading DNA in Genes."

YOU WRITE

Write your **descriptive title** at the top of the first page, centered, and bold typestyle. Leave a space below it and then continue.

Non-Descriptive Titles

You should *not* utilize trademarks or non-descriptive names that do not describe the subject matter of your invention. An example of an unacceptable invention title is "Jack's Great Doohicky"—this title simply does not describe what the invention is. Also, don't make the mistake of entitling your invention with a trendy description, such as "Zippy-Fast Zipper Opener." In our example, we would not want to call it "The Light Whammer," which may be used as its trademark. You want your description to be accurate, straightforward, and literally describe your invention.

Page 1

Lighted Hammer Apparatus

Background of the Invention

Field of the Invention

[0001] This invention relates to tools commonly used by carpenters, carpet layers, picture framers, mechanics, surgeons, archaeologists, jewelers and various other individuals. More specifically, this invention relates to the use of various types of striking tools, such as but not limited to hammers in low light applications or delicate operations requiring accurate hammer strikes.

Description of the Related Art

[0002] Any discussion of the prior art throughout the specification should in no way be considered as an admission that such prior art is widely known or forms part of common general knowledge in the field.

[0003] Hammers are commonly used throughout many trades to affix tacks, nails, staples and other fasteners in a variety of applications such as carpentry, cabinetmaking, art frame assembly, carpet laying, the automotive trade, and so on. Hammers are also commonly used in a number of more delicate operations by jewelers when faceting gems, by craftsmen and sculptors in conjunction with chisels, archaeologists and surgeons, such as in osteopathic and cranial applications. At times the various users must use the hammer in a low light application, which can be challenging, whether it is a tack or nail being set in a dark corner, a jeweler faceting a precious stone, a sculptor chiseling in a particularly tight space, a surgeon attempting to split a rib, or whatever the application may be. Even in adequately lighted conditions, the hammer itself may cause a shadow to be projected upon the surface that the hammer is intended to strike. Typical methodologies for using a

Page 2

hammer in a low-light application is to place lighting nearby thus shedding light on the fastener being set or nailed, a chisel being employed or any number of other skilled operations. An alternate method is to have an individual hold a flashlight or to place it nearby atop a desk, table or otherwise, and propping it up so the light rays to will illuminate the desired nailing area or item to be struck. It's understandable that both of these methods are inadequate at times due to the fact that the user may block the light that is shed from a lamp or flashlight, or pointing it in a tight space may just be physically impractical. The use of a flashlight such as the SNAKE-LIGHT® manufactured by BLACK & DECKER CORPORATION may help, however if this type of light is used, it too may be impractical to attach or point towards the desired surface or item being struck by the hammer. Obviously it is impractical to tote lamps and flashlights around to each and every construction site, archeological dig or household task. Nor is it practical to hire an extra employee solely for holding a flashlight or lamp. Whatever method is used, the inability to see or aim the hammerhead at the fastener, chisel, tool or other surface that is to be struck increases the frequency of injured fingers, results in a loss of productivity in a work environment, damage to the item boing worked upon, or may result in a great deal of frustration in a do-it-yourself home environment. In either case, the frequency of finger injury is substantially increased as well as the requirement for re-work, re-fastening and so on. In addition, the use of hammers in delicate matters (regardless of whether or not it is a low light application) such as faceting gems, carving sculptures, extracting dinosaur bones or the use in medical osteopathic applications is critical. While injured fingers from a miss hit may not be an issue in these applications, far worse scenarios may prevail. That is, a miss hit by a sculptor or gemologist may completely destroy the commercial value of the sculpture or the gem. A miss hit by an archeologist may destroy a 500,000 year old fossil. Likewise a miss hit from a surgeon may

Page 3

cause death of the patient or severe injury. As illustrated in U.S. Patent 5,628,556 Hrabar, the use of an internal light source coupled with a hollowed out tube provides an illuminated surface for a nut driver. This method serves its purpose most effectively, but it would be impractical to do with a hollowed out hammer. A means to accurately aim the hammer would be valuable to these trades and many others.

Brief Summary of the Invention

[0004] The lighted hammer apparatus of the present invention overcomes the problems associated with prior art. An extra lamp or flashlight is not required. Nor is any additional helper needed in order to safely set or nail a fastener in a low-light application, or use a chisel on a delicate matter such as a valuable sculpture, faceting a gem, archaeological or osteopathic applications. The illuminating light of the present invention is positioned within a lighted hammerhead specially designed to withstand multiple strikes to its impact face. The hammerhead cap is made of a high clarity material so that it projects a light beam directly on the fastener being set and nailed and the surrounding area, the chisel being tapped, the fossil being uncovered, the bone to be split in the operating room, or upon any other item or surface intended to be struck by the hammer. Regardless how tight the space may be the present invention projects the beam of light toward the desired surface or element being struck without concern of being blocked by the user or other nearby objects. The secondary benefit to the illuminated hammerhead is that it also serves at a sighting and aiming means for the user regardless of ambient light conditions. It is clearly understood that some skill is required to use a hammer, and with small brads and tacks it is even more difficult. When turned on, the light shining through the illuminating hammerhead on the present invention projects a light beam outward from the hammerhead, similar to that of a flashlight, which makes it easier for the user to hit

Page 4

the nail, chisel, surface or otherwise. The user uses this central focal point to first aim the hammerhead, and second, to hit the targeted item or surface. It may be equated to the use of some of the laser sighting devices used on military weaponry, except that the light source in the present invention is internal, literally inside the hammerhead which is being aimed, whereas lasers used on military weaponry is typically external to a gun or cannon barrel. Of course, the projected light can be focused into a highly concentrated beam, or be set as a broad-beam of light illuminating a wider area. This unique aiming ability may be extremely important to the jeweler, sculptor, archaeologist, or doctor, thus avoiding a miss hit that may make a valuable gem stone or sculptured masterpiece worthless, a fossil damaged forever, or a miss hit that could cause severe or fatal injury. The lighted hammerhead accomplishes this by using a hammerhead cap that is made of an extremely hard, durable clear material with a Fresnel lens effect that helps direct the internal light directly out of the impact face on the hammerhead. The light source may be of any variety of bulbs, but preferably is that of a fiber optic light emitting element located behind the striking surface of the hammerhead, proximal to the hammer handle. The light emitting element is connected to a switch or rheostat via wiring connected to a power source. The switch is preferably located immediate to the hammer's handle in the proximate location of the pointing finger, so that it may be activated, much like a trigger is activated on a drill motor, however, a switch can reasonably be located anywhere on the hammer. The lighting means may be electrically powered by an AC or DC power source, located within, or external to the hammer. An externally located power source would be connected to the lighting means by one or more electrical wires. The objectives of the present invention are to illustrate the following: 1) a hammer with a means to illuminate a surface that will be struck by the hammer; 2) a means of using and/or aiming the lighted hammer to illuminate a fastener, a sur-

Page 5

face, or other type of tool; 3) a means of manufacturing an illuminating hammer, and; 4) a special composition of material that provides the hardness and durability for use as a lens with the clarity and optics for use as a light focusing unit. Furthermore, it is an object of this application to illustrate the preferred embodiments and broadly state the methodologies that may be used in order to light up a low-light application, or aim the hammer in a variety of applications.

Brief Description of the Drawings

[0005] Fig. 1 is a side perspective view of the present invention illustrating the components that make up the preferred embodiments.

[0006] Fig. 2 is a side perspective side view of the present invention showing the internal components and how the present invention functions.

[0007] Fig. 3 is a rear perspective view of the unique hammerhead cap.

[0008] Fig. 4 is a side view showing how the internal wiring of the present invention may be positioned in the hammerhead.

[0009] Fig. 5 is a perspective view showing how the elements of Fig. 4 may be efficaciously assembled into an illuminated hammer.

[0010] Fig. 6 is a perspective view showing the concentrated beam of the illuminated light when placed in use.

[0011] Fig. 7 is a perspective view of a variation of the present invention.

Detailed Description of the Invention

[0012] **Overview:** In Fig. 1, hammer 10 has a handle component 20, with an internal battery power source 30, a switch mechanism 40, a hammerhead 50 and an illuminating hammerhead cap 70, which is securely affixed to hammerhead 50, which contains a lighting element therein (not shown). The high impact illuminating

Page 6

hammerhead cap 70 as disclosed herein is easily adapted to any number of ham-mer styles, such as but not limited to hammers with peen heads (such as illustrated in Fig. 1), but may include opposing claws, mallets and hatchets and any number of technical varieties used in the various trades. It will become easy to see that with lit-tle modification any hammer or striking tool may be adapted with an appropriate illu-minating hammerhead cap, small or large, making it suitable for any particular industry or use.

[0013] The switch mechanism of the present invention may be toggled or trig-ger-activated as described herein. Other types of switch mechanisms, including automated switches may be suitable to power on and off the illuminating light, how-ever we are illustrating only the preferred embodiment of a trigger activated switch which is intuitive to use and helps conserve energy and battery life.

[0014] It should be understood that the use of a hammerhead cap is the pre-ferred embodiment as it is not only replaceable thereby extending the life of the tool, but a user may also be able to use a variety of striking devices suitable for a partic-ular application, any of which will have the same luminary effects as described herein. A solid hammerhead made of the same durable, clear plastic may serve the appropriate purpose, but without the versatility of using a hammerhead cap.

[0015] **Illuminating Assembly:** In Fig. 2, hammer 10 has a handle 20, which has a hollowed base 24 where power source 30 is shown in the form of two bat-teries 32a and 32b inserted into power source 30 by opening end cap 22 and thus securing them in place, much like batteries would be inserted into a flashlight. Bat-teries 32a and 32b along with connectors 34a and 34b at the top, are housed in an internal casing (see Fig. 6) that together comprise power source 30. Power source connectors 34a and 34b are connected to wires 36a and 36b respectively, whereas

Page 7

wire 36a connects to off/on switch 40 (preferably a trigger switch as illustrated), which may be easily switched on or off by the user's pointing finger since the lower grip portion 26 of handle 20 is grasped by the user in a conventional manner, thus activating switch 40 in much the same manner as a user would activate a drill motor. Wire 36a then continues upward through the upper hollowed portion 28 of handle 20 alongside wire 36b, whereby the pair is then threaded upwards through hammerhead 50. Hammerhead 50 consists of an upper hammerhead plane 52, a central opening 54, which opening securely accepts uppermost tapered end 29 of handle 20, a front end 56 and an opposing peen ball 58. Securely attached to front portion 56 is illuminating hammerhead cap 70. Front portion 56 may be threaded either externally as illustrated 60, or internally in the same general location, in order to screw on illuminating hammerhead cap 70. Wires 36a and 36b are threaded through upper hollowed portion 28 and then feeds through a hollowed portion 62 of hammerhead 50, which wire ends connect to a light element 38 such as a fiber optic light emitting element, said light element 38 positioned at a cen-tralized point internal to hammerhead cap 70. When switch 40 is activated, power from power source 30 provides electricity through wires 36a and 36b thereby light-ing light element 38, thus illuminating hammerhead cap 70 and because of the Fresnel effect focuses a cast light L forward in a concentrated beam and out through impact face 72. The intention of this figure is to illustrate a preferred methodology of casting an illuminating light directly out of a hammerhead and out through its impact face. The unique means of casting light through an illuminated head centralizes the light in a desire area, upon a desired surface, or upon a desired fastener or element. Any number of combination of light emitting elements or hammerheads may be used, all of which would be considered under the scope of the present invention.

Page 8

[0016] **Hammerhead Cap:** In Fig. 3 illuminating hammerhead cap 70 is comprised of an impact face 72 (opposite side), an internally threaded base 74, which terminates in a flat butt end 76 (shaded), an internal reflective surface 78 in the form of any number of typical Fresnel lens configurations that may concentrate the light emitted from light element 38 (not shown) into a light beam in a preferred direction and level of concentration as it passes out through impact face 72. Hammerhead cap 70 may be made from a high clarity, durable plastic material that will withstand the required impact and shall also provide the emitted light to easily shine out through hammerhead cap 70 and impact face 72 as desired. The internal reflective surface 78 may be configured with any number of Fresnel lens configurations in order to concentrate the light beam towards impact face 72 as may be desired, including the ability to focus an even tighter, brighter (or at times darker) ray of light centered in the middle of the beam, which ray serves as a means to aim the hammer when striking a fastener, chisel or otherwise. Being made of a durable high impact polycarbonate such as a Lexan® (a trademark of General Electric Corp.), which is widely known for its unique combination of high impact strength and thermoformability makes it suitable for this application. Lexan also has superior optics, light transmittance (clarity). A polycarbonate material may also be blended with a UV inhibitor such as an HALS (hindered amine light stabilizer) to help protect against degradation in outdoor environments subject to increased light exposure. Furthermore a material such as GE's Lexan Margard provides an even greater superior impact resistance with the advantage of an abrasion resisting state-of-the-art coating. A Shore D durometer rating of at least 65 is desirable for most applications, however, a Shore A 85 to 95 rating may be desirable for applications suitable for certain applications where extreme hardness and impact strength may not be desirable but a certain amount of flexibility, or softness, may be.

Page 9

Table A. Material Hardness Scale

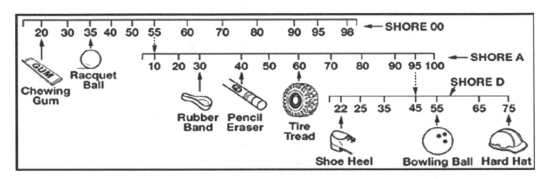

[0017] **Method of Manufacture:** In Fig. 4, hammerhead 150 is constructed to include an annular slot 152 (shown in phantom with dashed lines) extending from edge 154 of hammerhead 150 and faces the handle 120 (see Fig. 5). The width of slot 152 is such that it will receive handle wall 122 (See Fig. 5) of hammer handle 120 and is a thickness that approximates, and can be inserted into, annular slot 152. The depth of slot 152 is such that a hammer handle can be inserted sufficiently to provide the structural integrity needed when a handle and hammerhead are joined. Prior to molding the hammerhead 150, an electric light transmission cable (ELTC) 130 is inserted into the mold during the production cycle. As the mold is filled with the material comprising the ELTC 130 is held in the mold so that the light output end 132 is located in the center line location 134 of the finished hammerhead 150.

[0018] Figure 5 shows how the hammerhead 150 may be joined to the hammer handle 120 during the production cycle. A suitable connector 140 is used to connect the ELTC 130 contained in hammerhead 150 with the cable 124 that was previously assembled within hammer handle 120. A variety of means can be used to attach the hammerhead 150 to handle 120, including but not limited to a friction wedge, adhesives, mechanical fasteners, snap-fit details, rotational locking details, mating thread details located within the slot 152 and hammer handle 120.

Page 10

[0019] **Method of Use:** In Fig. 6, from hammerhead cap 70, and out through impact face 72, a concentrated light beam B illuminates a circular area C on surface S and directly upon and around nail N, which is to be set therein. In the center of light beam B is a center focal beam F that shines a tighter light ray upon nail N. This center focal beam serves as an aiming means so the user may more accurately strike the nail with hammerhead impact face 72.

[0020] **Variations:** In Fig. 7, hammer 210 works in much the same manner as the hammer in Fig. 1 with a handle component 220, an internal battery power source 230 accessible by opening end cap 222, a switch mechanism 240, a hammerhead 250 but without an illuminating hammerhead cap. In this variation of the present invention, hammerhead 250 has a chisel point 252 on one end and a pick 254 on the opposing end with internal light elements 238a and 238b lighting up either end of hammerhead 250. Light beams are thereby cast in much the same manner as the previous illuminated hammer illustrated in Figs. 1, 2 and 6, except that the light will not be cast from an impact surface, but will be cast from frontal areas 272a and 272b. The result is substantially the same, depending upon the internal Fresnel configurations located at 274a and 274b respectively. Unlike hammer 10 in Figs. 1 and 2, if a user of hammer 210 wished to replace the chisel or the pick end, he would have to replace the entire hammerhead 250, which may or may not include the lighted elements.

[0021] The spirit of the present invention provides a breadth of scope that includes all methods of lighting a hammerhead and focusing a concentrated beam onto a striking surface, element or another tool. Any variation on the theme and methodology of accomplishing the same that are not described herein would be considered under the scope of the present invention.

You can find very good suggestions as to what title you should use by reading the titles of granted patents that are similar to your technology. Of course, such titles may require some modification.

CROSS-REFERENCE TO RELATED APPLICATIONS

YOU WRITE

If this is your first patent application, you will not include the *Cross-Reference to Related Applications* paragraph. Just like in our Illuminated Hammer example, it is not included.

Under 35 U.S.C. §120, a subsequent permanent patent application is entitled to the benefit of the filing date of a prior application only if "it contains or is amended to contain a *specific reference* to the earlier filed application." It should be noted that a provisional application cannot cross-reference to a prior application. Nor would the first application have this requirement. Thus, if you're writing a provisional application, or this is your first on the subject matter, this section is not required.

When filing your permanent patent application, you must, therefore, reference the previously filed provisional patent application in order to claim its priority date. This may be particularly important if you have preserved international filing rights with a prior provisional patent application. Your attorney will automatically do this for you when converting your provisional patent application into a permanent application.

This rule applies to all subsequent patent application filings related to an original application, should there be a previous priority date you need to preserve. Examples include a continuation-in-part or divisional application.

Bear in mind that at times, you may NOT want to reference a previous patent application, which allows you to claim a new priority date. This may be desirable when filing new, related inventive matter. If you have several patents blanketing an invention, a more recent filing date extends the life of the patent protection on a product, which may also extend the life of a related license agreement. However, this should only be done in situations where you are confident that no prior art has been created before the filing date of the newly filed patent application.

For example, you may have related subject matter that can be filed as a continuation-in-part on an existing pending application, or you may file a new patent application altogether. Your decision should be based upon several factors. First, the one-year rule; second, the vulnerability of someone else perhaps filing similar subject matter during the existing patent application's pendency; and third, whether or not the subject matter has been publicly disclosed. A public disclosure will eliminate your ability to file a PCT (see page 161) if you elected to record a more recent filing date. Whatever you might be thinking, make sure you consult your attorney and discuss your options.

Thus, Rule 78(a) specifically defines what the "specific reference to the earlier filed application" must contain. The specific reference must contain the serial number, filing date, and relationship of the applications. The relationship of the applications would be "provisional patent application," "continuation," "continuation-in-part," or "divisional."

An example of an acceptable cross-reference is:

I hereby claim benefit under Title 35, United States Code, Section 120 of United States patent application Serial Number _____ filed _____ (hereinafter "Prior Application"). This application is a _____ [Legal Relationship] of the Prior Application. The Prior Application is currently pending. The Prior Application is hereby incorporated by reference into this application.

Again, this element is only a part of the permanent patent application. If you are writing a provisional application, skip this section.

STATEMENT REGARDING FEDERALLY SPONSORED RESEARCH OR DEVELOPMENT

If your invention was created as part of a federally sponsored research or development program, you will have to disclose the contract or grant number within the patent application. An acceptable illustration of a proper notice is:

YOU WRITE

The U.S. Government has a paid-up license in this invention and the right in limited circumstances to require the patent owner to license others on reasonable terms as provided for by the terms of Contract No. _____ awarded by the _____ [Agency].

This clause is quite uncommon and rarely used. Unless you are doing some sort of work for the government or some agency, you'll never need to insert this clause. Most likely you'll not include it in your provisional applications.

Insert the *Statement Regarding Federally Sponsored Research or Development* paragraph only if your patent application is part of a federally funded project—which MOST applications are not. We excluded it in the Illuminated Hammer application.

BACKGROUND OF THE INVENTION

This is where you begin writing about the history of your invention and its patentable subject matter. All patent applications must include this section. This is where the storytelling begins, the first part of the timetable. In this section, you will discuss the field of your invention and discuss previous and present-day products, and others that may not be closely related, but may be considered prior art.

The purpose of the Background of the Invention section is to provide a brief history and foundation for qualifying and substantiating your invention. The Background of the Invention section typically has two subcategories: the *Field of the Invention,* and *Description of the Related Art.* Each subsection is typically one paragraph long. (Refer to the example of this section on page 128–130.)

Field of the Invention

The Field of the Invention subsection of your patent application should provide an overview of your invention's field and the *purpose* of your invention—what problem or problems it attempts to solve and/or the benefits it provides.

You don't have to write in this subsection title, but including it at the beginning of the paragraph, as we did in our Illuminated Hammer patent application, makes it easier to write and follow the application.

Then, in your first sentence be specific and describe the field of your invention with words such as ours, "This invention relates to . . . "

As with the title of your invention, the field of the invention should provide the reader a clear understanding of what your invention is related to and what you are attempting to accomplish. The Field of the Invention should include information about the subject matter the invention relates to (hammers, flashlights, etc.). The Field of the Invention portion is usually only one or two sentences long.

Description of the Related Art

The Description of the Related Art subsection should identify the conventional products (prior art) in the industry and the problems associated with these products (more specifically the problems that your invention solves). In a storytelling sequence, try to discuss some of the older products first, leading up to the more recent ones.

ILLUMINATED HAMMER

In our patent application we broadened the scope of the use of the invention in the **Field of the Invention** subsection by using terminology that does not confine it to only common household or commercial tools.

YOU WRITE

Start the **Field of the Invention** subsection similarly to our example with the words, "The present invention relates to the field of _____. More specifically, it relates to _____" (be specific as to the manner in which your invention is used). You may also say, "The present invention provides (or improves on) the _____" (then explain).

However, remember that all statements that you make in your patent application may be used against you by the USPTO and later in a patent infringement lawsuit. If you admit something is prior art, the USPTO examiner may use this against you. Also, if you do not state the truth 100 percent, the validity of your patent may be questioned. Only make statements in this section that you know are 100 percent true. Don't guess or speculate. Your description of related art should clearly describe not only your technology, but the technology that others have invented that is closely related to yours. If it's related, it should be described fully. If you discuss prior art, make sure it is actually prior art that you simply cannot avoid with a non-analogous argument.

It is also helpful to include an introductory statement in the Description of the Related Art section that ensures the reader understands you are not admitting the references cited are actually prior art. Below is exemplary language to utilize in the beginning of the Description of the Related Art section:

Any discussion of the prior art throughout the specification should in no way be considered as an admission that such prior art is widely known or forms part of common general knowledge in the field.

Prior art may be comprised of existing or previous products that have been marketed. It may also include any patents that you have located that are similar in structure to your invention. With those patented products you plan to cite, you'll want to reference the patent number. For example, "The hammer in U.S. Patent 3,303,605 shows the use of . . . "

This is an important part of your application, because what you include here sets up the rest of the document to disclose how your present invention solves these prior art problems. Make sure you sufficiently describe the known problems with prior art that relate to your invention's solution. Not now, but in the subsequent sections, you will describe how your invention solves these prior art problems.

Another important consideration in this section is to make sure you have cited all prior art related to all of your patentable subject matter, whether that is a device, an associated method of use, a process of making it, or the chemical composition of the materials comprising it. In other words, if you're going to be claiming three or four types of patentable subject matter, you should cite the prior art related to all three or four.

ILLUMINATED HAMMER

In our patent application we discuss the Description of the Related Art subsection by referring to common, prior art hammers in general. The only reference we cite is the lighted nut driver invented by Kristin Hrabar. We cite that this invention works effectively with a nut driver, but cannot be applied to a hammer.

There are two commonly used ways to cite the prior art problems. Either describe the problems associated with any one prior art item as you discuss it, or at the end of the subsection, dedicate one paragraph summarizing "all the various problems associated with prior art."

This is particularly important if there is an existing patent that has some similarities to your invention, but may inadequately overcome the prior art problems. You want to be able to describe specifically what the potentially conflicting patent does and why—so you can prove that your inventive matter is useful, unique, and non-obvious. You want to describe exactly how your inventive matter is superior over this similar patent's subject matter. When the examiner reads your qualified, written statement describing how your invention overcomes the competing patent's inadequacies, and does not challenge it, in essence the examiner is agreeing with you. This is a good position to be in.

In contrast, if the examiner finds the subject matter in his or her search effort after the permanent patent application has been submitted, you've got a much tougher argument to overcome. Typically the examiner would cite that your invention is not patentable over the prior art, thus obviating your claims. In order to overcome this objection, your attorney will have to prepare a response, which could be costly and time consuming, and submit it to the examiner. Worse yet, this may not be accepted.

You will list closely related patents in your Information Disclosure Statement (IDS), and then include copies of those patents with your application. You're better off citing potentially objectionable patents in your patent application through your IDS, instead of having the examiner do it.

The Description of the Related Art subsection may be fairly short—only four to five sentences—or long, up to fifteen to twenty sentences. Again, what is most important is to broadly describe the prior art in a truthful manner, and the problems associated with it.

If you're struggling with prior art products to compare, you've probably not done enough research or you are not thinking in the right vein of thought. No matter how unique an invention may be, there is invariably some prior art. It is extremely rare that none exists. Even Edison's light bulb had prior art—regular candles, candles in illuminated glass lanterns, and prior art light bulbs that burned out in a matter of minutes or hours. Here are a few examples of the prior art associated with some exceptionally unique inventions:

INFORMATION DISCLOSURE STATEMENT (IDS)

An Information Disclosure Statement is a list of all patents, publications, U.S. applications, or other information submitted for consideration by the Office in a permanent patent application filed under 35 U.S.C. § 111(a), to comply with applicant's duty to submit to the USPTO information which is material to patentability of the invention claimed. See www.patentwriter.com for a sample IDS form.

- Television: prior art includes radios and the cinema.

- Transistor: prior art includes the cathode ray tube.

- Reading DNA in genes: prior art includes blood sampling, fingerprints.

- The cell phone: prior art includes the cordless telephone, walkie-talkies.

- The Internet: prior art includes local networks, teletype, and telephone.

- One-click, shopping-cart Internet business methodologies: prior art includes shopping at the mall, mail order catalogs, newspaper and magazine ads.

Do you see how virtually every product has some prior art it relates to? So must your invention.

If your invention is not related to any prior art, but is truly a completely new, never-before-conceived-of concept or process, then you'll want to consult with your patent attorney to determine how to approach this situation.

BRIEF SUMMARY OF THE INVENTION

The Brief Summary of the Invention section should describe in broad terms the overall structure, functionality, and operation of your invention. More specifically, this summary should describe the objectives of your invention, which are in essence the solution to problems associated with the prior art you've just discussed. (Refer to page 130–132 of the inset.)

The discussion of the overall structure should include a summary description of your invention in big-picture terms, and citing of the patentable subject matter. This structure will be similar to what will eventually be written in the Abstract of the Disclosure section discussed later, which is a part of the permanent application.

This section may be as short as four to five sentences or as long as twenty or so. Usually you will begin with comments that describe the benefits of your invention and how they overcome the problems associated with prior art. Specific sizes and shapes are not usually relevant to this discussion and may restrict the subject matter. The exception to this rule would be if your invention has certain specific tolerances or dimensions that are essential to its problem-solving function. In such a case, you'll want to cite the scope of those tolerances, dimensions, and so on.

ARGUE AFTER A FIRST OFFICE ACTION— NOT BEFORE

When drafting a patent application, it is tempting to put "arguments" as to the patentability of your invention before even getting an office action from the USPTO examiner. You may accidentally limit the scope of your invention or worse, create a validity issue. It is prudent to wait for the first office action before making arguments for the patentability of your invention.

YOU WRITE

The **Brief Summary of the Invention** section starts with "The present invention overcomes the problems associated with prior art. The present invention does this by . . . " (then, describe how it does).

**ILLUMINATED
HAMMER**

In the **Brief Summary of
the Invention** in our patent
application, we discuss the
benefits of being able to see
the area or the tool being
struck without the
requirement of additional
light, but we also describe
a secondary benefit of
having a central focal
point that can be used
to "aim the hammer."

**YOU
WRITE**

In the **Brief Summary of
the Invention** section,
define your objectives. For
example, "The objectives of
the present invention are: 1)
to overcome the _____
(refer to the product's
patentable subject matter);
2) to improve the _____
(refer to the method of use
patentable subject matter),
and; 3) to provide a superior
_____ (refer to the
superior manufacturing
process)," and so on.

For example, let's say you developed a plastic film heat-sealing process. You may say something like, "The plastic film produced under the process of the present invention is ideally 460 degrees, with a tolerance of ± 20 degrees." Thus, your invention is going to be restricted to that temperature range. But again, you do this only if that specific temperature range is the reason why your heat-sealing process works so well in the first place. Otherwise, you'd want to state the phenomena much more broadly, something like, "The plastic film produced under the process of the present invention is then heat sealed" and you avoid citing temperatures altogether.

In our example we say:

The lighted hammer apparatus of the present invention overcomes the problems associated with prior art. An extra lamp or flashlight is not required. Nor is any additional helper needed in order to safely set or nail a fastener in a low-light application . . . and so on. . . .

We continue with:

The illuminating light of the present invention is positioned within a lighted hammerhead specially designed to withstand multiple strikes to its impact face. The hammerhead cap is made of a high clarity material so that it projects a light beam directly on the fastener being set and nailed and the surrounding area . . . and so on. . . .

Once you've illustrated how your invention solves the problems of prior art, you should list the objectives of your invention so that a reader will understand what you are attempting to accomplish. By listing these objectives, you are literally making a list of the patentable matter, just like you outlined in your checklist (Chapter 5) and wrote in your claim statement (Chapter 6). If you have three patentable subject matters, you'll have at least three objectives.

Throughout the discussion of the objectives of your invention, you should identify all the unique features and benefits you're declaring. By doing so, you'll be building a convincing argument supporting your invention's usefulness and value.

BRIEF DESCRIPTION OF THE DRAWINGS

The purpose of this section is to identify what each drawing represents to assist the reader in understanding your invention and how it works. Each drawing is described in a single sentence and a single paragraph. If you have five drawings, then you'll have five, one-sentence paragraphs in this section. See page 132 for this section of our Illuminated Hammer sample application. The corresponding drawings (Fig. 1, Fig. 2, etc.) appear below, and on page 146.

This is by far the easiest section to do. What most experienced patent-drafting professionals do is make a list of the anticipated drawings in what they believe will be the correct storytelling sequence. Usually, you want to start with the basic views of your invention and then, in the proper sequence, show how it works.

It is important you take a little time to prepare this sequence. You don't want to jump around trying to describe how your invention works. By preparing this preliminary list of drawings, you are organizing the content you'll be writing in the next section. The content in the next section will reference the details of the drawings and describe how your invention works.

Once the list of the drawings has been completed, the professional may later go back over them and make any adjustments or changes to the one-sentence descriptions. On the computer, it's a simple task.

Likewise, you may find that while writing the next section, there is an additional drawing that should be included that was not on the original list. Thus, it is a simple task to insert it in the correct order.

ILLUMINATED HAMMER

In the **Brief Summary of the Invention** in our patent, we listed the primary objectives as a lighted hammer, a method of use, a process of manufacturing it, and a specified material (or composition of matter). In other words, four different patentable subject matters.

A NOTE ABOUT FIGURES

Although it is the book's style to use the word "Figure" when referencing drawings, when discussing illustrations *that are used in actual patent applications,* these will be referred to as "Fig." as is the style in patent applications.

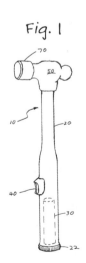

Fig. 1

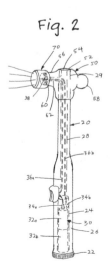

Fig. 2

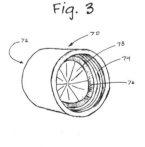

Fig. 3

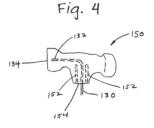

Fig. 4

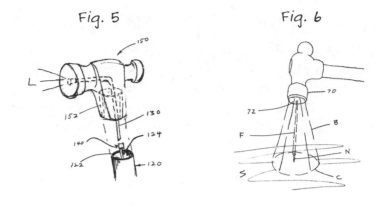

Fig. 5

Fig. 6

Fig. 7

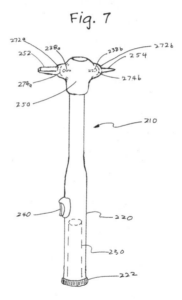

YOU WRITE

In the **Brief Description of the Drawings** section, make a list of the drawings you plan to use in your application. Put them in the sequential order you'll use to describe your invention and all the inventive subject matter you plan to claim (per your claim statement). You can make changes later.

YOU WRITE

IMPORTANT. Before you begin the **Detailed Description of the Drawings** section, review Chapter 9 and determine which kinds of drawings would best illustrate your inventive subject matter. Learn and apply the proper use of numbering and lead lines.

Each one-sentence description of a drawing should provide a direct statement regarding what type of figure it is (perspective, side view, plan view, and so on), and what the subject matter shown in the figure is about.

These drawings (in Figs. 1–7) are actual hand and computer sketches developed along with the application. As you can see, it is not necessary to have professional drawings. It is only necessary to include enough drawings to fully show the details of your invention.

Review the examples in our patent application and you'll see the simple format that's used. In the next chapter, you'll learn all about the various types of drawings you can use in your applications.

DETAILED DESCRIPTION OF THE INVENTION

Next to the claims section, this is a most important part of your patent application. There are a few reasons why. First, you must adequately describe your invention and its functionality (the specifications). If you fall short, the examiner may cite that there is no supportive reference to the patentable subject matter you claim.

Oh, you might show it in the drawing, but if you don't reference it, it becomes irrelevant. You might even cite it in this section, but if you don't describe exactly how it works, it'll be rejected. The solution to this problem usually results in a CIP or settling with a patent application that can't be fixed. (Turn to page 132 for this portion of our sample Illuminated Hammer application.)

Second, if you inadequately describe your invention in your provisional patent application, it begs the question, "What *really* is the inventive matter?" Your application could easily be vulnerable to legal scrutiny should a challenge arise. And you'll have a hard case to prove!

With all that said, follow our instructions and you should be able to describe your invention, accurately define the patentable subject matter, and set up the application to support the broadest possible claims.

Under 35 U.S.C. §112, an inventor must adequately set forth and fully describe three items within the Detailed Description of the Invention:

(1) the invention (the **description** requirement);
(2) the manner and process of making and using the invention (the **enablement** requirement); and
(3) the best mode contemplated by the inventor of carrying out his or her invention (the **best mode** requirement).

Just like the Background of the Invention and the Brief Summary of the Invention is written in a sequential storytelling mode, so is this section. While you are writing this section, you will be describing and referencing the drawings, which you would have laid out in the previous section, beginning with simple views showing how the invention looks, then into more descriptive views illustrating how it works.

Also, you must describe all patentable subject matter you plan to claim in sufficient detail. You must show how it works and why it is unique, useful, and valuable. For instance, after defining your invention as a device,

ILLUMINATED HAMMER

In the **Brief Description of the Drawings** in our patent application, the first one-sentence paragraph is, "Fig. 1 is a side perspective view of the present invention illustrating the components that make up the preferred embodiments." We accomplish two things here. First, we show the invention in its complete form and secondly, we show the "best mode."

POSSESSION OF INVENTION

The written description must disclose that the inventor was in possession of the invention at the date of filing the application.

UNDUE EXPERIMENTATION

The written description must disclose the invention so that one skilled in the art would not have to perform undue experimentation to construct and operate your invention.

you may show its method of use and how it is made, providing they're also considered novel.

Last, you will want to sufficiently describe the variations on your invention in order to blanket potential design-arounds. We'll also show you some shortcuts you can use to combine related subject matter and variations into a single phrase and avoid this problem. First, let's review the legal requirements of this section.

Description

The first paragraph of 35 U.S.C. §112 provides that the "specification shall contain a written description of the invention." The description requirement's purposes are to assure that the applicant was in full possession of the claimed subject matter on the application filing date, and to allow other inventors to develop and obtain patent protection for later improvements and subservient inventions that build on the applicant's teachings.

Enablement

Since 1790, patent laws have required inventors to set forth, in a patent specification, sufficient information to enable a person skilled in the relevant art to make and use their inventions.

The invention that must be enabled is that defined by the particular claims of the patent application. An enabling disclosure is all that is required. The applicant need not describe actual details of the embodiments or show examples. In fact, an applicant need not have reduced the invention to practice prior to filing a patent application. However, he or she must sufficiently describe how those embodiments function as it pertains to the inventive matter.

For example, when describing how the lighting device in our example works, it is not necessary to illustrate and describe technical details about light bulbs, switches, wiring, and so on. What is important is to provide an example of where they may be located and how they function together as a unit. By doing so, we are *enabling* the reader to understand the invention and the patentable subject matter we claim.

Best Mode

The last phrase of the first paragraph of 35 U.S.C. §112 provides that the specification "shall set forth the best mode contemplated by the inventor

---

BEST MODE TIP

Some inventors attempt to hide the best mode of their invention (best structure, best method of producing, and best method of using). Failure to disclose the best mode can result in the invalidity of your patent.

of carrying out his invention." The purpose of the best mode requirement is to prevent inventors from enjoying patent protection without disclosing to the public the preferred embodiments of the invention.

The best mode is commonly referred to as the "preferred embodiments" and is a chief component in the U.S. Government's pact to give you a legal monopoly (in the form of patent protection) in exchange for sharing with industry the state-of-the-art. A patent may actually be invalidated if this has been inaccurately disclosed.

Best mode is a subjective standard, as it focuses upon the best mode as contemplated by the inventor at the time the patent application is filed—not the objective best mode of the invention. In addition, if the inventor determines a better mode of carrying out the invention after a patent application is filed, there is no requirement for him or her to amend the patent application.

Writing the Detailed Description

This is the most difficult part of a patent application to write. When writing this section, try to be alone where you will not be interrupted, and try to have plenty of time to devote to writing and drawing. It is no small challenge to adequately describe in writing the structural elements relevant to patentable subject matter, their functions, and the operations.

It makes little sense to spend fifteen minutes a day trying to write this section, and taking weeks to do it, when you can most likely sit in a quiet room and finish it in a matter of a few hours. You'll invariably need to go back over your draft, compare it to the drawings you've prepared, and then modify, edit, and expand on the content. You'd be wise to prepare a complete draft in one or two sittings and then take your time in the coming days to perfect it.

There is good news, however, because the more experience you have writing this section, the easier it gets. In fact, after one or two patent application drafts, it becomes a relatively easy process. You may also consider using the *PatentWizard* software, should you feel you need assistance (see page 226).

Often it is helpful to first prepare the patent drawings you plan to utilize in the patent application. As you go, you may add certain elements to the drawings or add new ones altogether should it be necessary. You will then describe each figure as it illustrates a particular structure, function, or

DESIGN-AROUNDS

Experienced inventors know the value of writing patent applications. It is usually during the writing of this section that other designs, functions, and alternatives are found. If you plan to have a successful patent, you must make sure the scope of yours covers worthy design-arounds.

DISCLOSE IT ALL

Include more than one embodiment. U.S. patent law requires an inventor to describe only one embodiment of his or her invention. When doing so, the claims of the patent are typically not to be construed as being limited to that sole embodiment unless the inventor indicates otherwise. However, it is still prudent to describe more than one embodiment of your invention if you are aware of alternatives. As always, be careful not to limit the scope of your claims to the embodiments disclosed in your patent application.

ORGANIZATIONAL TIP

One trick you can use to help organize the written copy corresponding to your drawings is to copy the list you prepared under Brief Description of the Drawings and paste it into the section titled Detailed Description of the Invention. Then attach the subheadings to the list of figures and begin the detailed description. By doing this, you're explaining your invention in the correct sequential order.

process. As discussed in Chapter 9, Drawings, you should number the main components and sub-components according to standard practice and keep the application organized.

With experience, you may elect to do the drawings as you write the application. Generally speaking, this takes more skill and frankly, it also requires a highly visual mind. When taking this approach, you should still prepare a list of the drawings you anticipate in their sequential order as stated in the section Brief Description of the Drawings. Then, one by one, prepare the drawings while you write the corresponding descriptions.

When writing the Detailed Description of the Invention, you will disclose the invention in an organized manner. You may want to utilize subheadings to assist you in writing this section. It is not mandatory, but is a good way to help organize your drawings and written material. For example, you can start with the heading "Overview" and work through the inventive subject matter you plan to claim. These headings will correspond to the drawing list you've previously prepared.

For example, in our Illuminated Hammer apparatus application, we elected to use the following subheadings:

1. *Overview* (Fig. 1).
2. *Illuminating Assembly* (Fig. 2).
3. *Hammerhead cap* (Fig. 3).
4. *Method of manufacture* (Figs. 4 & 5).
5. *Method of use* (Fig. 6).
6. *Variations* (Fig. 7).

Overview

Under the first subheading, (Overview), you discuss the overall purpose and structure of the invention. Then, proceed as you describe the various components in more detail and all the inventive matter you plan to include.

As you write this section, you describe the invention in broad terms and refer to the numbered components as outlined in Chapter 9. Begin with the lowest number 10 to refer to the entire invention, and then in numerical order, describe the other components. Put the numbers right in your sentences without parentheses or brackets. For instance, we begin this section in our application like this:

Fig. 1

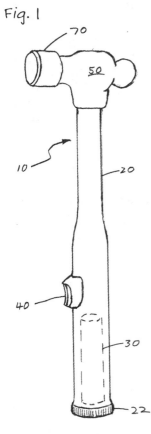

Overview: *In Fig. 1, hammer 10 has a handle component 20, with an internal battery power source 30, a switch mechanism 40, a hammerhead 50 and an illuminating hammerhead cap 70, which is securely affixed to hammerhead 50, which contains a lighting element therein (not shown).*

Just like we did, you should define all the major components in one single sentence. Each major component is an equal increment of 10 over a preceding component. The exception is that we jump from #50 on the hammerhead up to #70 on the hammerhead cap. We did this in anticipation of using a larger number of numbered references on the hammerhead.

In our example, we discuss the common hammer variety with a nail driving head and opposing peen head. Note that we have included a statement that many other forms of hammerheads may be used, such as an opposing claw, hatchet, and so on. We also expand on the various types of switches that may be used, power sources, and even hammerheads. In doing so, we are not limiting the scope of our invention. You should do the same in your application by using similar language.

However, when using these types of simplifying sentences, don't make the mistake of thinking you can use it to include specific inventive matter you've not described or illustrated. You must show enablement in order to claim the specific inventive matter.

For example, if the light switch in our hammer could be mounted on the handle or could be an automatic light-sensitive switch mounted above (or near) the illuminating light, you would have to show both varieties. If you didn't, there would be no reference to the placement and use of an automated switch and you couldn't claim it. You could say there may be "other switching mechanisms, placed elsewhere," however, if you did not

PRIOR ART DRAWINGS

If you plan to reference a prior art drawing, refer to Chapter 9 and see how these are drawn with reference letters instead of numbers. Usually prior art drawings are placed in this same paragraph, and generally precede the overview of your invention, Figure 1. If you are inserting a prior art drawing in this place, it is to illustrate a certain flaw. However, remember that anything you illustrate as "prior art" can be used against you by the examiner and an infringer attempting to invalidate your patent.

illustrate the specific invention of the "other switching mechanisms," you couldn't claim it.

What is important in this example is to use the broadening sentence terminology, and then have your attorney prepare a properly worded claim on the permanent patent application that blankets the use of all switch mechanisms. If, later on, a specific automated switch mechanism is invented, we may file a new patent application as an improvement thereon.

Illuminating Assembly

Under the second subheading (Illuminating Assembly), you would typically describe the detailed structure of your invention and how it works. In our application we begin like this:

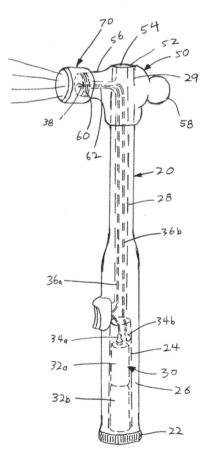

Fig. 2

Illuminating Assembly: In Fig. 2, hammer 10 has a handle 20, which has a hollowed base 24 where power source 30 is shown in the form of two batteries 32a and 32b inserted into power source 30 by opening end cap 22 and thus securing them in place, much like batteries would be inserted into a flashlight.

As you continue to read further in our application, you will see we have described the invention in detail and how the components and their elements interact. It is important to explain the details as we've illustrated them so the inventive subject matter is clearly understood. Too many inexperienced inventors attempt to write patent applications and illustrate their inventions as we did in Fig. 1, believing this one drawing is sufficient. It's not.

The second, more detailed paragraph ends with the following sentences:

The intention of this figure is to illustrate a preferred methodology of casting an illuminating light directly out of a hammerhead and out through its impact face. The unique means of casting light through an illuminated head centralizes the light in a desired area, upon a desired surface, or upon a desired fastener or element. Any number of combination of light emitting elements or hammerheads may be used, all of which would be considered under the scope of the present invention.

This kind of language is commonly used throughout patent applications to let the reader know that the scope of your invention covers all hammers in use (at least all those you can imagine!). Learn to spot these opportunities to state the broader concept by adding these kinds of statements. It will save you a lot of time and eliminate the need to describe and draw every last variation.

Hammerhead Cap

Under the third subheading (Hammerhead Cap), we discuss one of the invention's components in greater detail. You, too, should discuss individual inventive structures in your application in the subsequent drawings. The hammerhead cap we illustrate in our invention is of a particular clear material, such as a polycarbonate plastic, with a Fresnel lens to create a focal area for the light. The writing goes something like this:

TERMINOLOGY

The terminology you utilize in describing your invention is very important. For example, using the terms "light bulb" to describe the illuminating assembly would be overly narrow and would possibly prevent protection for potential embodiments that do not utilize a light bulb (e.g. laser, fiber optics, light emitting diode).

Fig. 3

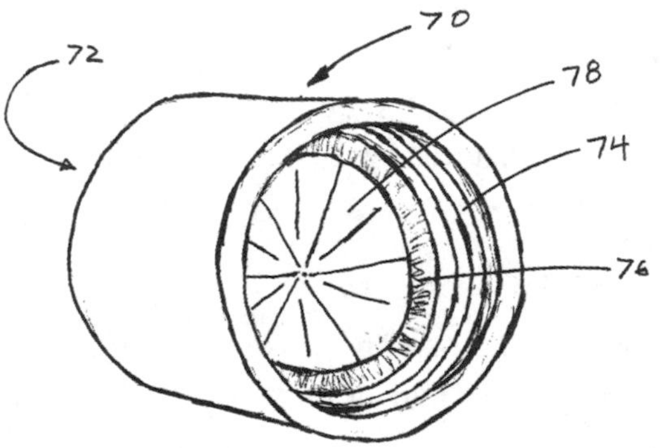

Hammerhead Cap: In Fig. 3 illuminating hammerhead cap 70 is comprised of an impact face 72 (opposite side), an internally threaded base 74, which terminates in a flat butt end 76 (shaded), an internal reflective surface 78 in the form of any number of typical Fresnel lens configurations that may concentrate the light emitted from light element 38 (not shown) into a light beam in a preferred direction and level of concentration as it passes out through impact face 72.

153

Here are a few words you can use in your invention's description. First, use words like "opposite side" to show a surface that cannot be seen in the illustration. It makes it a lot easier than preparing a separate drawing. Second, we use the word "shaded" to help illustrate a certain surface. Without the shading in the drawing, it would be a bit difficult to see that surface 76 is flat. Third, we use the words "not shown" as a means to describe a component of an invention that has been previously discussed but is not illustrated. It's obvious anyway.

After describing the hammerhead cap, you may expand on the inventive subject matter if desirable. For example, in our application, we cite a certain type of plastic material and rating. The reason we do this is that we intend to specify a chemical composition that composes the durable, yet high clarity, plastic we use to make a hammerhead cap. By doing this, we will be able to include a claim on our permanent patent application (to be filed up to a year later) with the material that best serves the purpose.

In this paragraph we have actually cited three different patentable subject matters:

1. A high clarity plastic hammerhead cap.

2. A plastic cap that has a Fresnel lens.

3. The ideal composition of the cap.

The broadest, of course, is the first one, for which we hope to secure patent protection, and that will be listed as an independent claim. The second one will be listed as a dependent claim on the first one. The third will be a dependent claim, as well referencing the first one. The third may also be written as a separate independent claim depending upon your intention and whether or not you believe there may be other applications for its use as some form of high impact lighting system. In such a case, this patent application would most likely be split into two separate applications after it is submitted to the USPTO. One would be for the Illuminated Hammer, the other for a high impact lighting device. This is called a CIP and is covered in Chapter 11. Don't concern yourself with this now, however.

Method of Manufacture

In the fourth subheading, we describe a cost-effective means of manufacturing the Illuminated Hammer. In your application you may describe a cost-effective means to make your invention or you may describe the unique manufacturing process of one or more of the various components. Here is how we approached the method of manufacturing:

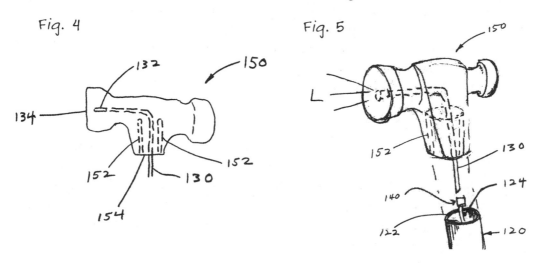

Fig. 4

Fig. 5

Method of Manufacture: *In Fig. 4, hammerhead 150 is constructed to include an annular slot (shown in phantom with dashed lines) 152 extending from the edge of the hammerhead that faces the handle 120 (see Fig. 5). The width of slot 152 is such that it will receive a hollow portion 122 (see Fig. 5) of hammer handle 120 that has a wall thickness that approximates and can be inserted into annular slot 152. The depth of slot 152 is such that a hammer handle can be inserted sufficiently to provide the structural integrity needed when a handle and hammerhead are joined. Prior to molding the hammerhead 150, an electric light transmission cable ("ELTC") 130 is inserted into the mold during the production cycle. As the mold is filled with the material comprising the ELTC 130 is held in the mold so that the light output end 132 is located in the center line location 134 of the finished hammerhead 150.*

In Figs. 4 and 5 we discuss how the hammerhead could be attached to the handle and how the illuminating means, the fiber optic lighting means, may be assembled inside the hammerhead. We also include some words you can use in your application. First, we use "(shown in phantom with dashed lines)" that helps us illustrate the invention. This is a simple way of illustrating subject matter without using a cross-sectional view or multiple drawings. Use them to your benefit whenever you can.

EXPAND SCOPE

If your patent application is a process application, you'll also want to claim "all products produced by the process." Make sure you cover this in your application. You can do this by adding a separate independent claim. There are two benefits to doing this. First, it expands your patent's scope. Second, if an import company is shipping an infringing product into the U.S., it is a lot easier to prove infringement with a representative product, than it is to go overseas into the plant and to prove it was using the infringing process.

WRITING TIP

The best way to write your variations quickly is to cut and paste a previous figure's description and then alter the description accordingly. For example, for Fig. 7, we cut and pasted the description in Fig. 2 and then changed or inserted certain words to describe a chisel point instead of a hammerhead cap. Note that the numbers also correspond with the preceding Fig. 2 except that they begin with "210" instead. From this perspective, it even makes it easy to renumber the various components and newly introduced variations.

Next, we use the parenthetic words "(see Fig. 5)" instead of explaining once again how the hammer handle is constructed. Last, we introduce the anagram "ELTC" as a means of simplifying the long term "electric light transmission cable." By doing this, we are able to quickly repeat the term that it applies to and speed up the written portion of the application. Frankly, the term "electric light transmission cable" is also a bit of a tongue twister.

Method of Use

We have actually been defining a method of use (a system) throughout the patent application. In this subheading and in Fig. 6, we show more specific details as to how it may be used in actual practice. By doing so, we will be claiming the method of use as an independent claim. We will also claim the method of use with a focal point as a dependent claim on our independent claim.

In our application we cite:

Fig. 6

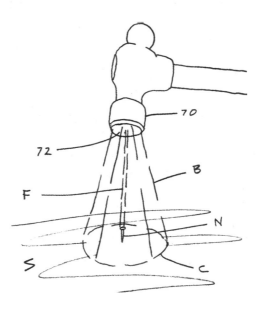

Method of Use: In Fig. 6, from hammerhead cap 70, and out through impact face 72, a concentrated light beam B illuminates a circular area C on surface S and directly upon and around nail N, which is to be set therein.

156

Simply put, we are describing how an individual will use the illuminated hammer to light a surface, find the object to be struck, and strike it thereon. You may use similar language when describing how your invention may be used by the user.

In the drawing, we also use *letters to illustrate a phenomenon.* For example, the letter B for "light beam," C for the "circular pattern," S for "surface," and N for "nail." Numbering may be used instead; however, it is also common to use letters when illustrated subject matter is not directly related to the invention.

Variations

In this last subheading, you will want to add any specific variations to your invention, should it be desirable. With our invention, we elected to add a variation on the head type. We could also consider adding the automated light switch as previously discussed.

Fig. 7

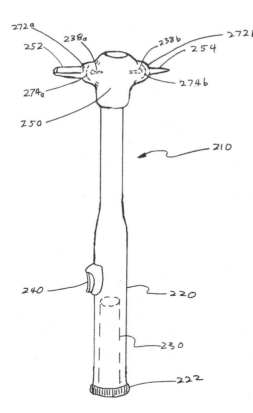

Variations: In Fig. 7, hammer 210 works in much the same manner as the hammer in Fig. 1 with a handle component 220, an internal battery power source 230 accessible by opening end cap 222, a switch mechanism 240, a hammerhead 250 but without an illuminating hammerhead cap. In this variation of the present invention, hammerhead 250 has a chisel point 252 on one end and a pick 254 on the opposing end with internal light elements 238a and 238b lighting up either end of hammerhead 250.

With our variation we have only changed the hammerhead to a chisel point and explained the benefits. In a way, we are using this explanation to declare "whatever type of hammerhead and/or point or tip" may be illuminated. At the end of this paragraph, we confirm the intention:

The spirit of the present invention provides a breadth of scope that includes all methods of lighting a hammerhead and focusing a concentrated beam onto a striking surface, element or another tool. Any variation on the theme and methodology of accomplishing the same that are not described herein would be considered under the scope of the present invention.

The good news is . . . at this point, you've just completed your provisional patent application!

CLAIMS

As discussed in depth in Chapter 6, the claims are the heart of a patent and should be prepared by a professional patent attorney when submitting the permanent patent application. Since you've already prepared your claims in suitable layperson terms as defined in your claim statement in Chapter 6, your work is essentially done.

Now that your application draft is complete, go over your claim statement once again to make sure you've adequately covered the patentable subject matter, and to make sure there are no additional claims you might want to add. You don't get a second chance to add additional inventive matter to a patent application later on. It would require filing a CIP or an entirely new application altogether. So take the time to review it now.

Again, a provisional patent application does not require claims. The exception would be if you would prefer to add one, as discussed in Chapter 5. If your objective is to file a permanent patent application instead of the provisional, then you'll be submitting your written draft and your claim statement to your patent attorney to prepare the legal claims and file the application.

ABSTRACT OF THE DISCLOSURE

An Abstract of the Disclosure is nothing more than a written summary of your invention. It goes on the cover page of the patent when issued. It is not required when filing a provisional patent application.

The purpose of the Abstract is to enable the USPTO and the public to quickly determine, from a cursory inspection, the nature and gist of the patent. The Abstract should contain the title of the invention, the purpose of the invention, and the general structure or function of the invention that makes the invention patentable.

The reason it's best to write the Abstract last is because once you've completed your patent application, it's much easier to summarize. In fact, you may use most of what you wrote in the section entitled Summary of the Invention as the content for writing your Abstract.

WRITING TIP

After you've written your disclosure, see if you can tell a family member the summary of your invention in just two or three sentences. This will be a test to help you craft a concise, accurate Abstract of your Disclosure.

If you are going to prepare the Abstract, cut and paste the Summary of the Invention section in a new paragraph and edit it into a concise and easy-to-read paragraph. You may also want to include language in the overview drawing (such as our Fig. 1) to broadly describe the invention. You must limit your Abstract to no more than 150 words. Should you have any doubt about doing this, let your patent attorney write it and submit it with you permanent patent application. It will only take him or her a few minutes to complete.

The Abstract in the illuminating light patent application, based upon the wording used in our Summary of the Invention and Overview (Detailed Description of the Drawings), would read something like this:

An illuminated light is positioned within a lighted hammerhead specially designed to withstand multiple strikes to its impact face. The hammerhead cap is made of a high clarity material so that it projects a light beam directly on the fastener being set and nailed and the surrounding area. The illuminated hammerhead also serves as a sighting and aiming means for the user regardless of ambient light conditions. The illuminated hammer has a handle component, a power source, a switch mechanism, a hammerhead and an illuminating hammerhead cap 70, which is securely affixed to the hammerhead 50, which contains a lighting element.

This abstract is 101 words long and describes the present invention succinctly.

FILING YOUR PATENT APPLICATIONS

With your patent application completed, you essentially have three filing options to consider. Hopefully you'll have established your patent strategy long before writing the application. Regardless of which option you utilize, it is still recommended to at least file your application through a qualified patent attorney to ensure proper filing.

Your three options are:

1. File a provisional patent application.

2. File a permanent patent application.

3. File a PCT patent application.

Filing Provisional Patent Applications

Filing a provisional patent application includes the cover sheet, the specifications, the drawings, and, of course, a check with the appropriate payment. Who files the application is a decision you'll have to make based on your experience and your budget. The order of desirability would be as follows:

1. **BEST: Your attorney reviews, your attorney files.** This way, your attorney can check for errors and omissions. Likewise, the attorney will be able to put the application on the firm's docket so that the attorney will automatically be advised of the statutory dates to file the permanent patent application.

2. **GOOD: Your attorney reviews, you file.** This may save a few dollars in handling fees for filing, but puts the burden of remembering the one-year statutory date for filing the permanent application on the inventor (you will have to personally monitor your upcoming one-year deadline). Send a copy of your application filing to your attorney, asking to have the one-year date docketed.

3. **ACCEPTABLE: No review, you file.** Experienced inventors primarily use this option. If you are inexperienced and have an extremely tight budget, it is generally considered best to at least file an application in this manner than do nothing at all. When using this approach, you may still want to notify your attorney of the filing and send a copy requesting him or her to put it on the firm's docket.

Experienced independent inventors and engineers file their own provisional patent applications all the time. Obviously, a company's policy may dictate otherwise, should you be one of its engineers or product developers. Before you consider this approach, it's best that you have a broad understanding of your industry, and experience in patent drafting.

Filing Permanent Applications

If your strategy is to file and prosecute a permanent patent application and skip the provisional phase, it will require a little extra time and, of course, added expense. Here are your options when filing the permanent application first:

1. **BEST: Your attorney prepares the claims and files.** This is the standard procedure all competent companies and inventors follow. All correspondence and office actions with the USPTO will then be directed to your attorney, since you'll be granting him or her power of attorney. Please understand that the patent attorney will have to review, modify, and possibly rewrite portions of your provisional patent application. The reason for this is that the patent attorney will be submitting the application as a legal document and will be responsible for the content contained within the application. No qualified patent attorney wants to file an untruthful and/or inadequate patent application with the USPTO, for obvious reasons.

2. **ACCEPTABLE: Your attorney prepares the claims, you file, your attorney responds to office actions.** This puts the application off the firm's docket, whereas you'll be notified of all correspondence and office actions with the USPTO. When received, you'll have your attorney prepare the proper responses. This can be a time-consuming process just to save a few dollars. It is rarely used.

3. **UNACCEPTABLE: You prepare the claims, you file, you prosecute.** It is never advisable for you to write claims, file, and prosecute your own permanent patent applications. The exception to the rule would be if you are an experienced patent attorney or patent agent in the field of your invention.

   Realistically speaking, there really is only one option—option 1. Remember, this is where the knowledge and expertise of your patent attorney is of critical importance in order to avoid a costly mistake.

WRITING TIP

Don't ever try to save a few dollars and try to write your own claims and prosecute them. This is really a costly mistake, since it will produce, at best, a patent with narrow scope that has little protection and is easy to design around.

Filing PCT Applications

This is similar to filing the permanent patent application, except that you will be filing a PCT (Patent Cooperation Treaty) application and designating those countries throughout the world in which you want to pursue patent protection. Thirty months later, you'll begin the national phase of filing within each country. If some of your contracting countries have not adapted their laws to reflect the changes to the PCT laws (effective April 1, 2002), you will only have twenty months to enter the national phase in those countries. However, even in those countries that have not adapted their laws, if you request an International Preliminary Examination within nineteen months of your filing date, you then have up to thirty months before entering the national phase for those countries as well.

Please keep in mind that you can file a PCT application within one year of your permanent application filing date, so the additional PCT filing fees are not required when you initially file the original United States patent application. Most inventors should consider filing only a provisional application or a permanent application, even if they desire to file in foreign countries.

This actually helps defray certain costs when filing, since it delays the national phase for at least thirty months in most countries. However, subsequently you'll be filing in the other countries and will be incurring those greater expenses afterward.

If this is your filing strategy, make sure you consult with your attorney and make sure his or her firm does the filing. Patent Cooperation Treaty applications are complex and require the services of a patent attorney experienced in filing PCT applications. There's no margin of error when pursuing this strategy.

SELF-FILING A PROVISIONAL PATENT APPLICATION

If you are filing your provisional patent application yourself, here's what you should do:

1. **Secure Papers.** Bind the written portion and the drawings together with a staple.

2. **Cover Sheet.** Completely fill out the provisional patent application cover sheet (it's self-explanatory and takes about five minutes). See www.patentwriter.com for this and other downloadable forms to file your provisional patent application.

3. **Check.** Include a check for the proper amount ($100 for small entities and $200 for large entities) written out to "Commissioner of Patents."

SAFETY FIRST

You have spent countless hours educating yourself and preparing a solid provisional application. Don't let "procedural" matters interfere with your legal rights. Have your patent attorney at least file your provisional patent application so you are assured of a filing date and an advance reminder to prepare a permanent application.

You can confirm the current USPTO filing fee for provisional applications by visiting the website www.patentwizard.com/fees.htm.

4. **Return Postcard.** Include a self-addressed return postcard that contains the following information on the back (the USPTO will stamp this card and return it to you for confirmation that your application has been filed):

- Name of first inventor.
- Title of patent application.
- Identification of all papers submitted including number of written pages, number of drawing sheets, filing fee, cover sheet, and other documents submitted.

5. **Express Mail.** Mail the above via U.S. Express Mail to:

Mail Stop Provisional Application
Commissioner for Patents
P.O. Box 1450
Alexandria, VA 22313-1450

On your U.S. Express Mail receipt, note the title of your invention and keep it for your records. You may immediately post "patent pending" once your local post office has taken your envelope. Make sure to visit www.patentwriter.com for updates on all of the above information.

In summary, using our Illuminated Hammer invention as a model, we've demonstrated how to write the various sections of a patent application. The Brief Description of the Drawings and the Detailed Description of the Invention introduce, and continually refer to, the drawings. It's rather easy to follow along with the description that refers to the drawings if they are well presented and clear. You're about to learn the secrets on how to easily and quickly prepare the drawings for your own patent application.

Drawings

Drawings representing your invention are easier to compile than you may think.

The drawings you use in your patent application don't have to be difficult to prepare. Surprisingly, patent drawings are not usually high-tech CAD-type drawings, but are more frequently viewed as simplistic sketches. Nevertheless, there are some basic rules you should apply so they are acceptable to the U.S. Patent Office, and so they will adequately depict your invention.

OVERVIEW

Under 35 U.S.C. §113, an inventor is required to furnish at least one drawing "where necessary for the understanding of the subject matter sought to be patented." With almost all inventions, this rule applies and most have at least three to six drawings or more to illustrate the patentable subject matter.

Mechanical inventions typically require drawings to illustrate the structure, function, and operation of the entire invention, including individual components. Method and process patents will require drawings to illustrate the overall connections, functionality, and operation (e.g. perspectives, flow charts, block diagrams).

DISCLOSE ALL EMBODIMENTS

Drawings are a great place to illustrate all alternative embodiments for your invention. Don't just disclose your preferred embodiment, but also disclose all possible variations of your invention.

If you utilize drawings within your patent application, you should make sure they fully disclose all of the invention's structure, particularly the patentable subject matter you wish to claim. In addition, the drawings should illustrate alternative embodiments to your invention, should you elect to claim them as well.

When making drawings of your inventions, use the same standards required in drawings approved for U.S. patent applications. If you have drafting or CAD experience, doing this is going to be fairly easy. If not, do your best to learn the following drawing types and reference number methodologies.

Keep in mind the drawings you submit to the USPTO, especially those you'll use on a provisional application, don't have to be professionally prepared—they only have to clearly and accurately depict the invention and the inventive matter. The USPTO will require the formal drawings to be prepared later, prior to issuance of the permanent patent.

When making drawings that illustrate how your invention works, you may use parts of drawings you've found in existing patents. You can even trace and resketch them, since there are no copyrights on patent documents.

DRAWING TYPES

There are several types of drawings you may use in your patent applications; however, there are only a few that represent the vast majority of inventions. In the pages that follow, we'll illustrate the most commonly used types.

Don't forget that your inventor's journal should be full of drawings that you have made throughout the invention development process. If possible, make the drawing entries in your journal similar to those we have illustrated so you'll get into the habit of preparing them according to USPTO requirements. Of course, make sure your journal is an approved one, meaning that the pages are pre-numbered and permanently bound so that material cannot be added or removed later on, and is therefore legally defensible.

Should your invention use more sophisticated drawing types such as electrical schematics, cross-sectional views, and so on, you're probably an engineer in the field and don't need any advice on how to prepare these drawings anyway.

Nevertheless, for most inventions, there are seven types of commonly used drawings in patent applications.

Plan Views

The plan view is typically seen from directly above an invention, in other words a "top view." It is one-dimensional and shows the basic layout, or plan, of an invention. Figure 9.1 is a plan view of a bookend. Note the three shade lines indicating a flat surface.

Perspective Views

These are by far the most commonly used drawings in patent applications. A standard three-dimensional drawing will show the invention and the inventive subject matter in a way that more closely resembles a real life perspective. It is not uncommon to have most of the drawings in an application be this type.

There are also variations on the perspective view such as "side perspective" or "top perspective," but it is not essential to be so specific. See Figure 9.2 for a perspective view of the bookend in Figure 9.1. In the Illuminated Hammer application, we used several perspective views.

Blown-Up Views

When a certain element in an invention is important, but may be of small size, it is best to illustrate it by using a blown-up view. In our Illuminated Hammer application, our Fig. 5 on page 155 is a blown-up view of the hammerhead cap in its entirety. Another common method of illustrating with a blown-up view is shown in

Figure 9.1

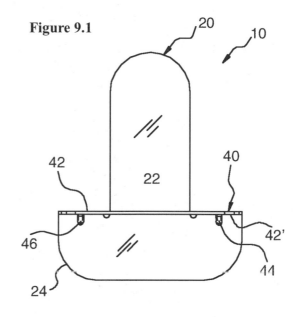

Figure 9.2

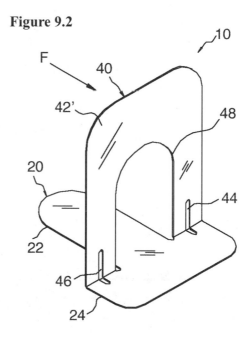

167

Figure 9.3

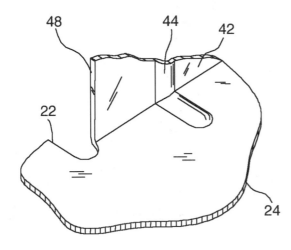

Figure 9.4

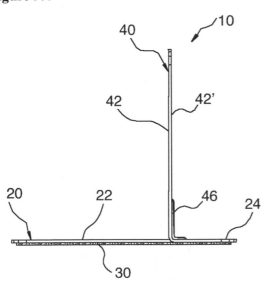

Figure 9.3 of the bookend on page 168. This is actually a "partial blown-up view" and may be labeled as such in your Brief Description of the Drawings.

Front, Side, and End Views

These may be helpful in illustrating the "head on" appearance of one or more of the one-dimensional views of the invention. They are not very commonly used in patent drawings, but are certainly one type of drawing that you can use to help illustrate your invention. Figure 9.4 is an end or side view of the bookend. In most cases, a perspective view can show all of the components and elements in a front, side, or end view, but with more clarity.

Exploded Views

This view may be important for illustrating the connection between invention components, while also illustrating each of the components individually.

It can be time consuming to make two different drawings and then try to explain how they fit together. An exploded view may accomplish that objective in a single sketch.

Exploded views may be used with a side view, end view, top or bottom view, or a perspective view, as we've applied to the bookend in Figure 9.5 on page 169. In the view of Figure 9.5, we're showing the felt pad base that would be affixed to the bookend itself. Note the speckled surface of the felt pad component. Exploded views are important to illustrate the interconnection and assembly of individual components.

Flow Charts

Flow charts are becoming more important, and typically illustrate how a methodology or system works. They may also be effectively used with computer- and software-related inventions. In Figure 9.6 on page 170, we illustrate a method of searching for data. (In the Illuminated Hammer application, we did not use a flow chart, since the method of use was explained in the body of the patent application and partially illustrated in Fig. 6 in Chapter 8 on page 156 in a perspective drawing.) When you use a flow chart in your methodologies, apply the correct forms of box styles and arrows indicating the various uses. Here is a list of the most common shapes and their use:

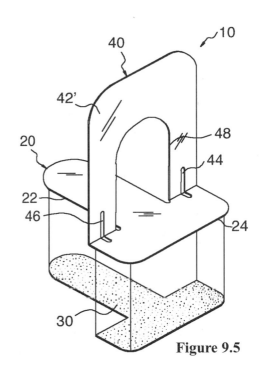

Figure 9.5

- **Oval-shaped boxes:** These are called "terminators" and are used for beginning and ending operations in a process or method.

- **Rectangular boxes:** These are used for intermediate operations that comprise the process.

- **Diamond-shaped boxes:** These are used to literally ask a question or to make a decision. In other words, they may illustrate whether or not a particular function has been fulfilled.

- **Arrowhead lines:** These will show the flow of the process. When leaving a diamond-shaped box, there will typically be two or more lines, thus illustrating with a "yes" or "no" whether or not a certain operation has been completed.

There are other box styles such as a parallelogram used for "data" or trapezoidal boxes that indicate a machine process that will be performed. For a more in-depth understanding of the various forms of boxes, symbols, and line types used in flow charts, see the information provided in the Resource List on page 228.

Block Diagrams

Block diagrams are useful for illustrating various electrical components

BENEFITS OF BLOCK DIAGRAMS

Block diagrams are excellent for illustrating the interconnection and communication between different types of well-known technology (e.g. a transmitter communicating with a receiver).

Figure 9.6

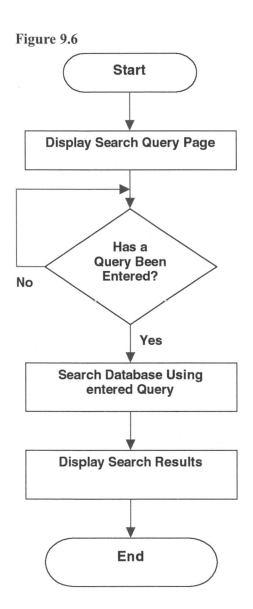

without having to illustrate physical wiring connections. These can be extremely useful when developing new electronic inventions where the specific wiring or electronic components need not be specified.

In Figure 9.7 on page 171, we show how a simple temperature control system may work on a pump. In such a block diagram, you are illustrating the interrelated functions in rectangular boxes. Each box is connected by lines.

You may use block diagrams to illustrate any number of inventions related to machinery, remote control devices, and even some business and training systems.

NUMBERING, LETTERING, AND LINE TYPES

Your drawings and the components will be referenced using a numerical format. By adopting the following formats and suggestions, you'll quickly become adept at patent drawing. Use the figures on the previous pages for reference. The following are guidelines for using reference numbers.

Figure Numbering

Designate a figure number for each drawing in your application. Keep the figure numbers in their proper sequence as you draw your sketches and describe the inventive matter. Do your figure numbering as we've illustrated in these figures. It is common to abbreviate the word figure with the abbreviation "Fig."

Lines with Arrows

Use a floating arrow-line (the number 10 in Figures 9.1, 9.2, 9.4, and 9.5) to reference the "invention in its entirety."

Use an arrow-line touching the edge of a main component to reference "the entire component" (for instance, 20 and 40 in the preceding Figures 9.1, 9.2, 9.4, and 9.5).

In the flow chart in Figure 9.6, lines with arrows represent the directional flow of the process operations.

Figure 9.7

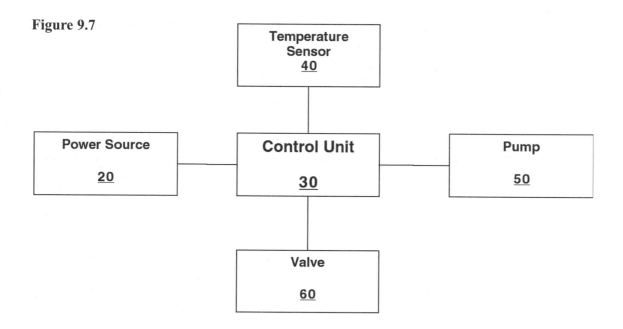

Lead Lines

Use a solid lead line without an arrow to reference an individual element, part, or portion of a component as illustrated in Figures 9.1–9.5 (pages 167–169) and Figure 9.8 (page 172). In the block diagram in Figure 9.7, lead lines are used to show the connection between components. Notice how lead lines usually have a little curve to them, as depicted in Figures 9.1–9.5.

No Lead Line

A lead line may be eliminated by placing a number directly upon a designated component and underlining it. This is illustrated as 22 in Figure 9.1.

Numbering Sequence

When numbering components and elements, begin with the number 10, such as in Figure 9.1, to identify the invention in its "entirety" and then number each of the "main components" of your invention in a sequence of 10s (20, 30, 40, 50 . . .).

After all of the main components have been numbered, you then should number the "sub-components," or elements, of each of the main components using only even numbers, based upon corresponding main component numbers. For example, if a main component was a base 20, the sub-components may be numbered as a first portion 22 and a second portion 24, which together form base 20. See Figure 9.1. It is often useful to create an "index table" to keep track of your numbered elements.

The reason for using only even numbers is that it is easy to overlook a particular element when writing your patent application and describing a particular figure. Thus, when you are writing the description in the Detailed Description of the Drawings section, and realize you've forgotten an element, you may go back and put in a number in the correct sequence. This, of course, would be an odd number.

For example, if you had forgotten to define an element that should have been described between two other elements 38 and 40, you can add that new element as number 39. This is illustrated in Figure 9.8. This way, the numbering is sequential as the drawing is being described. Otherwise, you would have to erase much of the numbering you've already completed, and readjust all your written material accordingly, just to keep them in a proper sequence. Or, you'd be inserting a much larger number (depending upon where you left off), and putting the numbering out of order in your drawings, which is not considered protocol.

Using a Prime with Numbers

When your invention has a right and a left side, or a front and a back, use the same number for components on each side. On one side, add an apostrophe (') after the number (referred to as a "prime"). For example, the number on a front side of a component is 42, and the number for the rear side of the component would be 42' as illustrated in Figure 9.4.

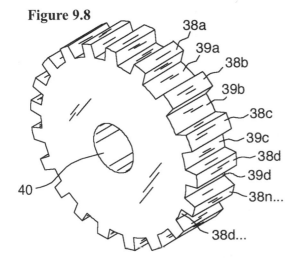

Figure 9.8

38a
39a
38b
39b
38c
39c
38d
39d
38n...
38d...
40

Similar and Same Elements

You can also add letters to the numbers when referring to similar elements. For instance, the teeth in a gear could be illustrated as 38a, 38b, 38c, 38d, and so on, as shown in Figure 9.8. This is also sometimes used in the same manner as the preceding group by identifying a left side component and a right side component (or front and back and so on). Primes are used to identify features on opposite sides of a part when they cannot be seen simultaneously, but the use of alpha characters to a part number are used when the same element is seen multiple times at once—like the teeth of the gears.

Letter Usage

Letters may also be used to illustrate forces and common phenomena. They are helpful as a means to repeat an action, as we've shown with the letter "F" in Figure 9.2.

For example, "F" is used for "force," "A" may be used for "air flow," and "V" may be used for "vacuum."

Letters are also commonly used to designate attributes on prior art drawings. For instance, in a prior art bookend drawing, "B" may refer to a base, "U" to an upright, and "R" to a reinforcing element (not illustrated).

In Figs. 5 and 6 of the Illuminated Hammer application on pages 155 and 156, we use several letters to illustrate light beams, focal points, and prior art elements.

Electrical and Mechanical Engineering

If you're an engineer developing electronic, electrical, or mechanical innovations, you'll use the obvious symbols for things like electrical and electronic components, building materials, pneumatics, hydraulics, and so on.

Your patent examiner is going to be an engineer knowledgeable in the field, and will readily understand the various components and symbols. It is unnecessary for us to duplicate the knowledge you, as an engineer, have already gained.

Shading

Shading can be important for illustrating the surface structure, material

COMMON COPYING MISTAKE

It is important for self-drafters to utilize the drawings of prior art patents. However, do not make the mistake of copying one or more figures from prior art that is so similar that you may have to argue against later. You want your invention to have an appearance different from the prior art patents, so it is important not to directly copy the drawings of prior art patents.

DRAWING AID

The *Scientific Journal* (see Resource List) is one aid you can use to help prepare drawings. The measured grid pattern makes it easy to prepare perspective drawings as well as all others. You may copy the drawings on a light setting on a copier and the grid pattern in the background will fade out, thereby making it easy to use in your patent applications.

types, and other important features of your invention. Shading is comprised of one or more lines that illustrate the surface contour, structure, material type, and similar information.

Shading can be confusing initially, but after reviewing some of the prior art patents that relate to your invention, you should be able to determine the type of shading to use on your invention. Shading is typically not crucial for your drawings and, in many cases, not required to fully illustrate the structure of an invention. However, if you feel it will help one skilled in the art to understand your invention better, you should at least attempt to utilize shading to illustrate shapes and structures of your invention.

PREPARING YOUR DRAWINGS

HIRING A DRAFTSPERSON

If you do not feel qualified to prepare your patent drawings and would like to hire a draftsperson, you should hire someone who has had extensive experience with patent drawings (i.e. not just technical drawings). Your patent attorney can refer you to a qualified draftsperson.

By the time you are ready to prepare your drawings, you should have a good idea of how your invention works and the patentable subject matter you'll be describing in your patent application, which drives the claims. Before preparing your drawings, you should know the following basic acceptable standards.

Paper and Drawing Size

The typical paper size you'll use is $8\frac{1}{2}$" x 11", the same as for the written portion. You'll also maintain a 1-inch margin at the top, bottom, and sides. (You are actually allowed to have a right margin of $\frac{5}{8}$ inches and a bottom margin of $\frac{3}{8}$ inches, but it is easier and safer to just use 1-inch margins for the entire drawing page.) Thus, paper size restrictions tend to dictate the drawing size.

The key rule to follow is: Your drawings must adequately depict the inventive subject matter—in other words, the patentable subject matter that will determine the claims you'll be pursuing. To do this, it is best to prepare drawings of sufficient size and clarity. Generally speaking, you do not want any more than two drawings per page.

Drawing Sequence

A well-written patent application discusses the inventive matter in a logical sequence. When you're ready to write that application, it will be easy to determine this sequence.

Once you've prepared your claim statement from Chapter 6, you essentially have a template for writing your patent application and preparing the corresponding drawings. It's really that simple.

Your drawings don't really dictate the order in which they'll be placed in a patent application. The order in which you'll be discussing the subject matter to secure subsequent claims will determine the order.

If you're still wondering what drawings to include, you should first review your claim statement and then determine which ones will illustrate the inventive subject matter best. Start out by listing these drawings in order in the section Brief Description of the Drawings (see Chapter 8). You can always go back later on and makes changes in the order, or insert new drawings altogether.

Quality Requirements

As we've pointed out, the type and quality of drawings used in patent application are not nearly as technical as those drawings typically used in the field of engineering. You can say they are more on the level of drawings that have been prepared based on a ninth grade drafting class. Sometimes they're not even that sophisticated.

If you believe you cannot adequately prepare drawings that meet the minimal standard, then you'd best employ someone to prepare them for you. However, keep in mind once again, some of the drawings you'll use with your patent application may be deleted later on, or you may have to make certain corrections based on errors or unclear subject matter cited by an examiner.

From this perspective, you'd be wise to submit drawings that adequately depict your invention, but ones that are not necessarily considered formal drawings such as those required prior to issuance of a patent.

Formal Drawings

Formal drawings done by a professional draftsperson are required by the USPTO for permanent patent applications. However, formal drawings are *not* required for provisional applications. Even though formal drawings are not required for a provisional application, you should still attempt to include professional drawings with your provisional application to ensure your invention is properly disclosed.

Photographs and Computer-Generated Output

USPTO rules 81–88 prescribe formal requirements for drawings (37 C.F.R. §§1.81–1.88). The rules allow submission of color drawings or photographs only under limited circumstances (37 C.F.R. § 1.84[a][2], [b]). See USPTO MPEP § 608.02.

Photographs are commonly submitted with plant patents in order to show their particular appearance. It would be somewhat impractical to submit an artist's conception in their place.

Also, there are times during prosecution when the examiner may ask for additional information in order to better understand an invention. Photographs are sometimes used to clarify these matters.

If you believe you need to include color drawings or photographs within your patent application, you should consult with your patent attorney regarding how to do this.

We have discussed every element of drawing preparation that will help you professionally convert the hand sketches from your inventor journal to clear drawings and illustrations in your patent application. You don't need to worry about perfection. You only need to show and label the important parts of your drawings, so that someone reading your description will instantly be able to see and understand the components that you are describing. These well-labeled drawings will help you double check your written application to make sure that you have discussed every important feature of your invention, and will help streamline review of your patent application by your patent attorney.

Chapter 10

Design Patents

*A simple way to expand your patent coverage and, in many cases, a
simple method to prevent infringement.*

We have spent a significant amount of time discussing the preparation of a utility patent application (provisional and permanent).
There is another type of patent application you should consider if your
invention's appearance is appealing—a *design patent application.*

A design patent only protects the overall appearance of an invention
and is usually less desirable than a utility patent. A design protectable by
a design patent may consist of surface ornamentation, configuration, or a
combination of both. A product infringes upon a design patent if it so
resembles the patented design as to deceive the ordinary observer who
gives such attention as a purchaser usually gives. It should be noted that
an inventor may apply for both utility and design patent protection.

Design patents may provide broader protection than a utility patent if
the invention is within a crowded art or if the invention is a very simple
structure. The term of a design patent is fourteen years from the date of
issuance of the patent. There are no maintenance fees for a design patent.

ORNAMENTAL VERSUS FUNCTIONAL

To qualify for a design patent, the configuration of a useful object must
not be dictated solely by function. A design patent is granted for any new,

original, and ornamental design for an article of manufacture. To receive design patent protection, the same rules apply for novelty as for utility patents. It should be noted that the non-functional design of a container or of a product itself might also serve to indicate origin and hence constitute a protectable trademark, though recent court decisions have cast doubt on this overlap.

Figure 10.1 shows an illuminated lighted apparatus with an ergonomic handle design that is both ornamental as well as functional. Figures 10.2 and 10.3 depict two variations on the hammerhead design. These three figures would be considered alternative embodiments, and are suitable for us to use in our design patent application.

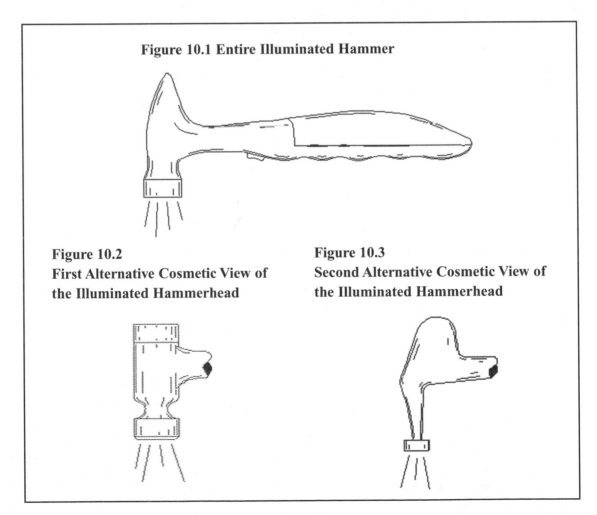

Figure 10.1 Entire Illuminated Hammer

**Figure 10.2
First Alternative Cosmetic View of
the Illuminated Hammerhead**

**Figure 10.3
Second Alternative Cosmetic View of
the Illuminated Hammerhead**

Therefore, if your invention's appearance is dictated solely by function, design patent protection may not be applicable. Consult with your patent attorney if you have questions as to whether you should seek utility and/or design patent protection.

WRITTEN DESCRIPTION

A design application does not have a detailed written description like a utility application does. The written description of a design patent application is basically comprised of:

- Cross Reference to Related Applications
- Federally Sponsored Research Statement
- Title
- Description of the Drawings
- Single Claim

The Cross Reference to Related Applications and Federally Sponsored Research Statement are the same as utilized within a utility application (Chapter 8). The Title and Description of the Drawings are also similar to that of a utility application. However, there is only one claim with a design application, and this claim is merely a broad statement identifying the drawings.

Description of the Drawings

Every design application has a Description of the Drawings section. This section is very simple and merely describes what view of your design a particular figure illustrates.

An example of a suitable figure description for a design patent application is:

FIG. 1 is an upper perspective view of a [INVENTION TITLE] showing my new design.

You should describe each drawing you attach to the application.

One Claim

A design patent contains only one claim. Full disclosure and definiteness of scope are provided primarily through the drawings.

179

An example of a suitable claim for a design patent application is:

I Claim: The ornamental design for a [INVENTION TITLE] as shown and described.

Nothing more is required for your claim section in a design application. As stated previously, the most important part of your design application is the drawings section.

DRAWINGS

The drawings of a design application are the most important part of this type of patent application. You should take special care to ensure that the drawings fully and accurately illustrate the ornamental features of your invention. You should closely review Chapter 9, which discusses preparing patent drawings. While design patent application drawings are similar to those of utility patent applications, there are stricter requirements relating to shading, which typically require the services of a qualified draftsperson.

Since the protection of your design patent will be dictated solely by the drawings you submit to the USPTO, it is extremely important to hire a professional draftsperson who has design patents experience. A qualified draftsperson can make sure you fulfill the USPTO requirements.

HIRE A DRAFTSPERSON

Many inventors attempt to file for design patent protection by preparing their own patent drawings. Unless your invention is very simple and you are a draftsperson yourself, it is strongly recommended to hire a professional draftsperson to prepare the drawings for a design patent application in order to meet Patent Office requirements.

Non-Essential Features

You should try to keep out non-essential and functional features of your invention if possible. For example, if your design is a unique eraser design attached to a pencil structure, you should not include the structure of the pencil within your drawings (alternatively, you may include the pencil by using dashed lines to draw it).

Figure Numbering and Views

Your drawings should fully illustrate the design of your invention by showing the design from various views. Each figure of the drawings should be numbered consecutively (e.g. Figure 1, Figure 2, Figure 3, etc.). You should not utilize reference numerals within your design patent application drawings like you would with a utility patent application (see Chapter 9). You will not be referencing elements in the written portion of the application.

Views typically included within design patent applications are:

- Upper/Lower Perspective Views (Three-Dimensional)

- Top/Bottom Views

- Front/Rear Views

- Right/Left Side Views

- Cross Sectional Views (where required to illustrate design features particularly difficult to see from normal drawings)

In our example, the following five figures represent the actual views and illustrations of a bogie wheel utilized within snowmobiles as illustrated in U.S. Patent No. D451,064. Notice that the drawing types follow the same format and description as those in a utility patent application.

Following in Figure 10.4 is a left side view of a bogie wheel.

Figure 10.4

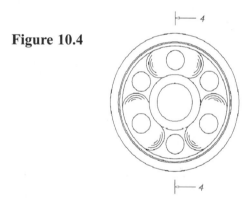

Next, in Figure 10.5, is a right side view of the same bogie wheel. Notice how the right side view is similar to the left side view.

Figure 10.5

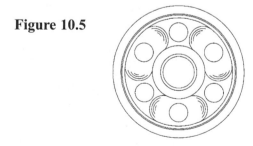

Figure 10.6 is an end view of the bogie wheel.

Figure 10.6

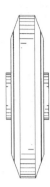

Figure 10.7 is a cross sectional view taken along line 4-4 of Figure 10.4.

Figure 10.7

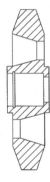

Figure 10.8 illustrates an upper perspective view of the bogie wheel. Of all the views, perhaps Figure 10.8 is the one that most clearly depicts the ornamental appearance. It allows us to envision the actual appearance as if we were viewing it in a real time perspective.

Figure 10.8

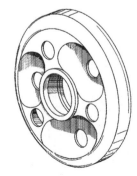

You can see from these drawings that various views were presented in order to illustrate the various visual attributes of the wheel. You will also note that these drawings include no functional details since functional attributes are not included in design patents.

These wheels can be large, small, thin or thick, be manufactured from wood, steel, or glass. Those functional attributes are unimportant in a design patent. However, regardless of the obviously missing manufacturing method, materials, size, or other functional attributes, the one important element of the design patent for this wheel remains: no other wheel may *look like* this bogie wheel when compared to any of these drawing perspectives.

Alternative Embodiments

Your design drawings should illustrate the unique ornamental structure—nothing more. However, you also should include various embodiments of your design that you have created. For instance, if you had considered a cosmetic design that included four large concave areas (rather than the three large concave areas around this bogie wheel), or if you had considered a design with eight or ten holes arranged around the center (rather than the six holes shown on these drawings), those alternative embodiments should also be shown by including the additional drawings.

We've shown you the third type of patent application that you may consider—the design patent. This patent will not disclose or claim any functionality whatsoever; it is used exclusively to patent the visual or aesthetic attributes of the invention. As you can see, by writing both a utility patent application as well as a design patent application on our Illuminated Hammer invention, a design patent can be used to lend an additional level of protection, beyond a utility patent.

Special Considerations

Once your patent application is written, there are several interesting strategies you may consider when filing and prosecuting.

By this time you have a firm grip on the whos, whys, whens, and hows of patent drafting, right? Well, just when you think you have everything under control, something else creeps into the formula and messes up everything.

Well, almost . . . but not quite.

Fortunately, since you are following *The Patent Writer* methodology, your margin of error will be substantially reduced. This is particularly true if you have an expert patent attorney on your team.

As you know, the patent process follows the requirements set by patent law, and sometimes, laws change. Actually, since the passage of the American Inventor Protection Act (AIPA) of 1999, the implementation of the United States Patent Office's 21st Century Strategic Plan, and certain court rulings, patent laws have gone through a metamorphosis.

Is it your job to become a patent legal expert? Of course not. It is your job to be an invention expert, however, and that means at least staying abreast of the shifting sands of patent law, and how that may effect your patent strategy and tactics.

We've outlined some of the more important considerations that you may want to keep in mind as you pursue your patent. As we've already said, patent laws tend to go through a transformation, so it's more important than ever to talk to your patent attorney if you believe any of these considerations apply to your current invention. You should also request your patent attorney to keep you informed of any new laws that may effect your new or pending application.

PUBLICATION OF APPLICATIONS

Beginning March 2001, the USPTO started publishing (publicly disclosing) utility patent applications (not design or provisional patent applications) filed on or after November 29, 2000. The rule requires the USPTO to publish the patent application not later than eighteen months after the filing date, except in cases where specific requests for non-publication are made, as is described later on. In exchange for the publication of a patent application, patentees may be able to obtain a reasonable royalty during the period beginning on the date of publication of the application by the USPTO and ending on the date the patent is issued ("provisional rights"). An application may also be published earlier than at the end of such eighteen-month period at the request of the applicant.

Don't mistake the publication of a patent application with the publication of an issued patent. Even though the patent application is published, the prosecution may take many more months or years before the patent is finally granted (the patent pending period). It's also important to understand that the statutory publication of a patent application is no guarantee that the patent will be granted—in fact, it may be denied by the USPTO after the application is published.

An application will not be published if an applicant makes a *Non-Publication Request* at the time of filing the application, certifying that the invention has not and will not be the subject of an application filed in another country, or under a multilateral international agreement, that requires eighteen-month publication. The Non-Publication Request may be withdrawn at anytime if an applicant decides to file in a foreign country.

The Non-Publication Request must be included with the patent application at the time of filing, and cannot be filed at a later date. Without the filing of a Non-Publication Request, a patent application will be automatically published by the USPTO unless the applicant expressly abandons the application. Ask your attorney about this request if it is of interest to you.

Published applications can be utilized by the USPTO to reject applications filed one year or more after the publication date thereof. However, published applications also fully disclose what an applicant is applying for, which may be undesirable if the applicant wants the subject matter to remain confidential. Finally, if a patent application is published, the USPTO charges a publication fee to the applicant if the application is later allowed.

PATENT COOPERATION TREATY (PCT)

Prior to the *Patent Cooperation Treaty* (*PCT*), when a United States applicant wanted to seek patent protection abroad, separate applications had to be filed in each country or region where patent protection was desired. Since foreign applications generally must be filed within one year after the first application anywhere pursuant to the Paris Convention, this imposed a severe financial burden, since the cost of foreign filings could not be deferred beyond that one-year period.

Effective January 24, 1978, however, the United States ratified the Patent Cooperation Treaty, which permits the filing of a single international application within a one-year period of filing a U.S. patent application. In the international application, any of the over one hundred PCT member countries or regions may be "designated." Most countries in which one might want patent protection are now PCT members, including the United States, Canada, Mexico, all European states, Japan, Australia, New Zealand, South Korea, China, Russia, Brazil, and many others. Note that there is no such thing as an "international patent"—the PCT provides an international patent application, but that international application still eventually leads to individual national patents.

By filing one international patent application under the PCT, you can simultaneously seek protection for an invention in each of a large number of countries (now more than one hundred) throughout the world. If you are a national or resident of a PCT Contracting State, you may file such an application and receive the following benefits.

- **Delay of Foreign Patent Costs.** One of the main advantages of the PCT is that it provides a significant opportunity to defer (not necessarily avoid) the major portion of foreign patenting costs. This provides much more time to assess or test the markets in various countries prior to committing to major patenting expenditures.

- **Preliminary Examination.** The PCT provides applicants with the option of obtaining (in addition to an international search report) an international preliminary examination report, providing information about the patentability of your invention—before you incur any costs associated with the patent-granting procedure in any of the countries in which you still wish to obtain patents.

- **Centralized Amendment.** The PCT also provides the possibility of complying with a number of formalities in a centralized manner when you prepare your application in accordance with the international standards effective under the PCT.

- **Fee Reductions.** The PCT provides significant fee reductions throughout the international and national phases of the procedure.

Filing a PCT First

One option is to file a PCT application prior to filing a United States patent application. A PCT application is filed with the World Intellectual Property Organization (WIPO) in Geneva, Switzerland. The PCT application generally goes through the same kind of examiner review as patents go through in the United States Patent and Trademark Office. The examiners will also conduct a prior art and novelty search for the PCT application.

At the time of filing a PCT, the patentee will designate the countries in which he or she ultimately wants to receive a patent. The PCT application will never mature into a "PCT patent." However, once the PCT application is allowed and published, the application will then enter what is referred to as the *national phase.* The application will be handed off to the patent office of each country designated at the time of filing of the PCT.

The examiners will review the PCT application, along with the international search report provided by WIPO. If the examiners in each designated country do not find a reason to not issue the patent, then they will send the patentee a notice of allowance, and require that the patent issue fee be paid. Thereafter, the patent will be issued by every designated country that finally approves the invention during this national phase of examination. The patentee will then be required to pay the appropriate patent issue fee to each country.

Your best defense against having the PCT patent application option slip out from under your patent strategy is to maintain close communications with your patent attorney.

FESTO

In the landmark *Festo* decision, the United States Supreme Court held that any "narrowing amendment" of a claim made to satisfy the Patent Act's requirements shall invoke "prosecution history estoppel."*

As an example, an original claim may have stated "A device comprised of two components joined together by use of a connection means" meaning that nails, glue, screws, bolts, or any other number of means could be used to join the two parts. However, the patent examiner may finally approve the claim once it is narrowed to specifically identify the means used to connect the parts, such as "A device comprised of two components joined together by use of two or more bolts running between and through the two separate components." It's easy to see that the changes narrowed the claim so that the patent specifically does not protect the invention if the two parts are joined by the use of glue.

With the amended claim, the patent cannot claim the benefit of the Doctrine of Equivalents (meaning that for all practical purposes, the connectors originally claimed in the previous example include nails and bolts and so forth, and any of the connectors or fasteners could effectively be used interchangeably).

Therefore, if the claim is amended, the inventor is prevented (estopped) from claiming that, although bolts are described in the approved claims, nails or glue could have been used and would have practically the same effect as if nails or glue were actually described in the claim.

Under this ruling, even an amendment to a claim element for the purpose of curing a problem of non-enablement, lack of description, or indefiniteness under Section 112 creates an estoppel, provided that the amendment "narrows" the claim element. In *Festo,* the Court also confirmed that arguments without amendment can create a rebuttable estoppel. Under *Festo,* "[a] patentee who narrows a claim as a condition for obtaining a patent disavows his claim to the broader subject matter."

In addition, *Festo* is not a complete bar of the Doctrine of Equivalents (see glossary) to the narrowed claim element. For example, equivalents that are unforeseeable at the time of the amendment may be protectable under the Doctrine of Equivalents. In addition, equivalents that were not surrendered by the narrowing amendment may be protectable under the Doctrine of Equivalents. In other words, only the subject matter surrendered

---

*Festo Corp. v. Shoketsu Kinzoku Kogyo Kabushiki Co., Ltd., 535 U.S. 722, 122 S.Ct. 1831 (2002).

by the narrowing amendment is subject to the *Festo* limitation. It is, therefore, important to draft claims that do not require significant or unnecessary amendments.

In summary, what does all this mean to the inventor and the inventor's business? Since we never recommend that you file your own claims, we put the burden on your patent attorney to ensure that your claims have power behind them, and that they are not carelessly written. Frankly, the best way to avoid any problem based on the *Festo* ruling is threefold:

1. Search the prior art thoroughly (e.g. patents, publications, products). After you have established what the prior art is, you can then determine what is patentable and tailor your claims to the same.

2. Before you write and file your patent application, prepare a thorough claim statement. This is your plan for ensuring that you'll have potentially broad scope.

3. Hire an expert patent attorney who is up to date on the laws. Don't settle for anything less! Your future may depend on it.

PHILLIPS v. AWH—CLAIM MEANING

The *Phillips v. AWH Corporation* case was decided July 2005 with significant impacts on the interpretation of claims. (A copy of the *Phillips* case is downloadable at www.patentwriter.com for your review.) The most significant aspect of the *Phillips* decision is that it focused the interpretation of the claims on the *intrinsic evidence* (i.e. the patent itself, statements made to the USPTO, and the cited prior art).

Claim Construction

The purpose of *claim construction* is to determine how a person of ordinary skill in the art of the invention's technology would understand the "ordinary meaning" of claim terms at the time the application was filed (i.e. not the inventor, not someone in a different industry, not someone with above-ordinary skill in the art).

Claim interpretation typically requires the courts to review two types of evidence—the patent itself, including its USPTO file history ("intrinsic record"); and evidence outside of the intrinsic record such as dictionaries, encyclopedias, and treatises ("extrinsic record"). Unfortunately, there has been significant confusion between the courts as to what methodology should be followed in interpreting claims.

"Ordinary Meaning" Before *Phillips*

Prior to July 12, 2005, the courts were split on how to balance the usage of the *intrinsic evidence* with the *extrinsic evidence.* There were two lines of cases prior to this time—the *Texas Digital* line of cases and the *Vitronics* line of cases.

The *Texas Digital* line of cases refers to those in which the courts first utilized extrinsic evidence to determine the "ordinary meaning" of the claim terms, and then considered the intrinsic evidence afterwards to determine if the inventor used the words of the claim in a manner "clearly inconsistent" with the ordinary meaning reflected by the extrinsic evidence (dictionary).

Hence, the courts in the *Texas Digital* line of cases adopted the ordinary meaning of a claim term as defined by extrinsic evidence *unless* the inventor provided clearly inconsistent meanings for the claim terms in the intrinsic record (lexicography). The reasoning in the *Texas Digital* cases is that by reviewing extrinsic evidence first, the court is less likely to read limitations from the specification into the claim.

The *Vitronics* line of cases refers to those in which the courts first utilized the intrinsic record to determine the ordinary meaning of the claim terms. While *Vitronics* acknowledged that judges are free to consult extrinsic evidence, the usage of extrinsic evidence could not be used to contradict any definition contained by a reading of the intrinsic evidence.

"Ordinary Meaning" After *Phillips*

On July 12, 2005, the U.S. Court of Appeals for the Federal Circuit issued its much anticipated opinion on claim construction in *Phillips v. AWH Corporation.* In *Phillips,* the Court provided additional guidance on how to properly interpret claim terms in a patent:

- If a patent only describes a single embodiment of the invention, the claims of the patent are typically not to be construed as being limited to that embodiment.

- Claim interpretation is focused upon how a person of ordinary skill in the art of the invention would understand the claim terms.

- The sequence of steps used by a judge in consulting intrinsic and extrinsic evidence is not important—what is important is for the court to attach the "appropriate weight" to those sources of evidence.

- While extrinsic evidence is still important, the intrinsic evidence is the most significant source for determining the meaning of claims.

- The intrinsic record typically will indicate whether the inventor is merely providing examples of his or her invention in the specification, or if the inventor intends for the claims to be limited to the specification.

- A court is not barred from considering extrinsic evidence as long as it is not used to contradict a claim meaning that is "unambiguous in light of the intrinsic evidence."

PHILLIPS POINTER

It may be beneficial to have your patent application reviewed by a person skilled in the art of your invention (under a confidentiality agreement, of course). This review ensures that your application is understood by one skilled in the art as you actually intended it to be understood! Ambiguities can be easily identified and corrected at this time—it is much harder to correct ambiguities after the application has been filed.

Here are some key points to follow in self-drafting your patent application in view of *Phillips:*

- **Understand the Terminology Used in Your "Art."** If you are going to write a patent that will be interpreted as one of ordinary skill in the art would understand the same, then it is important that you fully understand how people (e.g. engineers, technicians, other inventors) talk and use terms in the industry your invention is in. Consult technical dictionaries, patents, websites of companies, product manuals, articles on technologies, and other sources of information related to your invention that will provide guidance on how terms are used in the field of your invention.

- **Use Ordinary Terms.** Utilize only terms that have established ordinary meanings in the field of your invention. In other words, you should write your patent as one of ordinary skill in the art of your invention would understand the same. You should only utilize terms as they would be understood by one of ordinary skill in the art of your invention (e.g. terms as used in recent patents related to your invention, terms as used in technical dictionaries related to your technology).

- **Be Consistent with Terms.** It is important to utilize terms consistently between the claims and the specification. Any difference in usage may be utilized by a court to limit the scope of your claims.

- **Do Not Limit Ordinary Meaning.** We strongly advise against trying to be your own lexicographer when writing your patent applications. In other words, do not try to provide meanings to a term that already has an established ordinary meaning to one of ordinary skill in the art.

- **Do Not Limit to Embodiment Illustrated in Specification.** It is also important to ensure that one skilled in the art of your invention would not conclude that your invention is limited to the embodiment(s) shown in the specification. You should always utilize language such as "preferably" or "for example" to help avoid limiting your invention to a specific embodiment.

- **Diligently Watch Upcoming Court Decisions.** It is important to monitor all of the upcoming court opinions that will be issued interpreting *Phillips,* which you can use to assist you in the preparation of your patent applications. Make sure to visit www.patentwriter.com for updates on all court decisions related to the *Phillips* decision.

By utilizing words consistently that already have an established meaning in the field of your invention, you will be assured that if your patent ever has to be interpreted by a court of law, it will provide the meaning you intended—and not a meaning you did not intend or were unaware of!

LEGISLATIVE CHANGES TO PATENT LAW

Community, county, state, and federal laws are constantly changing. Over the past 200 years, the laws governing the United States Patent and Trademark Office have constantly been revised as well. Since the passage of the American Inventor Protection Act of 1999, a swell of proposed changes to patent law has mounted, with some of the very significant changes considered imminent.

It's advisable to meet periodically with your patent attorney during the development of your patent application so that you can remain abreast of any changes to law that could affect your patent strategy.

Here is a summary of proposed changes to the U.S. patent system that you should keep a close eye on (as always, visit www.patentwriter.com for the latest updates):

- **First-Inventor-to-File.** This proposal would change the U.S. patent system from a *first-to-invent* system to a *first-inventor-to-file* system. Under the new proposed system, in questions amongst two or more competing inventors, the patent would go to the inventor with the earliest effective patent filing date.

- **Treble Damages for Patent Infringement.** Currently, courts can allow up to treble damage for "willful" infringement. Under the proposed change, treble damages could be awarded only if the patent owner gave the defendant written notice of infringement containing specific information, or if the defendant copied the invention, and the defendant did not have a basis for a good faith belief that the patent was invalid, unenforceable, or not infringing. Lack of an opinion by an attorney would not create an inference of willfulness.

- **Post-Grant Opposition Proceedings.** Under the proposed legislation, anyone would be able to challenge the validity of a patent during a nine-month window of time following the date of patent grant.

- **Assignee Filing.** Current U.S. patent law requires the inventor to file the application even if the ownership rights are assigned to a company. Under the proposed legislation, a company that has been assigned the invention by the inventor will have the right to file the patent application, provided that the inventor is notified of the filing.

- **Publication of All Applications After Eighteen Months.** Currently, patent applicants can "opt out" of having their patent application published after pending in the USPTO for eighteen months by filing a Non-Publication Request. Under the proposed legislation, all patent applications will be published.

USPTO EXAMINATION REGULATIONS

There are several regulations that may affect the administration and processing of a patent application. The most commonly used regulations that you may use to your advantage are divisional applications, continuations, and continuations-in-part. These are all good to know so that you can make decisions later on that may actually improve the quality of your patent applications and patentable subject matter.

Why file one patent, when you can file two patents for the same price? The practice of filing an application for a continuation-in-part (CIP) or a divisional patent has long been a standard practice of companies that want to build a solid patent portfolio. We'll give you an overview of the CIP/divisional application process, but keep in mind that the following scenario is just one example. Every invention is different, and the patent strategies being pursued by inventors and patentees differ.

Also, more sophisticated patent strategies will ultimately require more investment in patent filing fees and attorney costs. Of course, the additional costs are well worth the additional patents and patent protection if it supports a patent strategy, but budgeting more money will be a requirement.

Finally, keep in mind that a patent strategy begins before you write your patent application. You will need to include various inventions in the initial patent, and filing a patent disclosing numerous inventions requires care. Therefore, before pursuing any CIP or divisional patent strategy, discuss the options with your patent attorney.

Divisional Applications

If an application claims more than one independent and distinct invention, the examiner may impose a restriction requirement. The applicant then elects one invention for prosecution in the original application and may file *divisional applications* claiming the other inventions. When pursuing this approach, you pay one fee and protect the integrity of the classification system in the USPTO.

Here is how this works. In our illuminated hammer patent, we have shown and described many different elements. To name a few: an illuminated hammer; a hammer handle that contains electronics and a power source, or connection to an external power source; and a striking surface crafted of a durable transparent material that allows light to be shown through it.

Now, the illuminated hammer is certainly novel and patentable. Therefore, a patent will be issued on the "Illuminated Hammer."

By including and describing in detail the hollow handle with electronics and/or connection to external power source, we are opening up the possibility of claiming an entirely different invention. The "electronics in the handle" invention may be applied to fishing rods, ski poles, walking canes, and so forth. As you can see, if we determined that the market for electronics in the handle was sufficiently large, and if the invention taken separately was determined to be novel and non-obvious (satisfied the patentability requirements), then we would be ready to file a divisional patent application on the "electronics in the handle" invention.

A divisional patent describes the condition where you would divide the current patent application into two or more separate patents. Here's how the process would go with our application:

SURPRISE

Now that we've fully described how divisional patent applications work, we'll put in a small wrinkle. The United States Patent Office continually reviews and considers legislation that would streamline office operations, and in fact, bring the USPTO processes more closely aligned with the patent application and examination processes used by other countries. One such consideration to standardize the treatment of disclosure involves the filing of multiple inventions. It is conceivable that they may be treated more like individual patent applications, or as CIPs, and the cost to file may be greatly increased. Thus, it's important to keep abreast of these kinds of rule changes to avoid any surprises.

- The patent examiner issues an office action that says we have disclosed multiple inventions, and asks us to elect one of the inventions for the present patent application.

- We elect the "Illuminated Hammer" application for the current patent.

- We take the description of the "electronics in the handle" invention, along with all drawings, background of the invention, and claims that we've included in the original patent application, and we file a new application limited to the electronics in the handle invention.

The advantage of this approach is that the original filing date of the first patent also becomes the filing date (priority date) for the second application. Also, rather than paying for two patent applications in the beginning, you have the advantage of waiting until the patent examiner asks you to elect one invention for the first patent. Since the first office action can take up to thirty months after filing your application, this means that you have delayed the payment for a second application by up to two-and-a-half years!

Another benefit is that if you determine during the prosecution phase that you have no intention of pursuing commercialization of the "electronics in the handle" patent, you can abandon it without ever making an investment in a separate application.

Continuation Applications

A *continuation application* is a second application for a patent filed by the same inventor and containing the *same disclosure* as a prior application (often times called the *parent application*). If the continuation patent application is filed during the pendency of the parent application (i.e. before issuance, abandonment, or termination of proceedings, including any appeal), and makes a specific reference to the parent, the continuation is entitled to the benefit of the parent application's filing date.

A continuation application can be used to secure further examination and to add claims after a final action by the examiner. A series of continuation applications (grandparent, parent, etc.) is possible, provided each application makes reference to all previous ones in the chain.

Continuation applications may be used in situations where the parent application has received a final rejection, but the applicant wants the ability to continue amending the claims freely. Continuation applications are also useful where an applicant is not satisfied with the initial patent protection granted by the USPTO and desires to seek broader protection through the filing of a second application.

Continuation-In-Part Applications (CIP)

A *continuation-in-part application* (*CIP*) is like a continuation application except that it adds some new subject matter to the original application. A CIP is entitled to the benefit of the parent application's filing date to the extent that the CIP contains common subject matter. A CIP may be used to add improvements developed after the filing date of the parent application.

A continuation-in-part is also similar to the divisional application. Just as a divisional patent application may result in a second patent (for a separate inventon), a CIP may also result in a second patent. However, the second patent that results from a CIP will often be an improvement on the first invention, rather than a separate invention.

For example, if, during the prosecution phase, we determined that a commercially sound opportunity existed for an illuminated hammer with interchangable hammerheads, *while the first patent is pending,* we could file a second patent application claiming the interchangeable illuminated hammerhead.

Again, we could claim the original patent filing date as priority; however, the priority date on the interchangable illuminated head would be the actual filing date of the CIP. Claiming priority of the earlier patent application prevents the earlier application from serving as prior art to the second CIP application. Otherwise, if the earlier application was prior art, the CIP would not be allowed. The CIP will nevertheless have a different filing and issue date.

Once again, it's vitally important that you keep in close communication with your patent attorney to be sure that you are making decisions based on the most currently applicable patent laws.

Continued Prosecution Applications (CPA)

A *continued prosecution application* (*CPA*) is either a continuation application or divisional application filed under 37 CFR 1.53(d). A CPA may be utilized only for utility, plant, or reissue applications. The CPA basi-

PAY NOW OR PAY LATER

You should keep in mind that filing a patent containing more than a single invention may save initial filing fees for multiple patent applications, but you will need to pay separate filing fees when you file your CIP or divisional application.

cally continues the prosecution of the patent application where the prosecution phase has come to an end.

CPAs may be a valuable tool when an applicant is having difficulties prosecuting a patent application, and has received a final rejection, thereby limiting his or her ability to amend the application.

Requests for Continued Examination (RCE)

Since 1999, 35 U.S.C. Section 132(b) directs the USPTO to adopt regulations allowing an applicant to request further examination. The purpose of "continued examination" is to give applicants an option of paying a fee and continuing prosecution, despite receipt of a final rejection.

Continued examination under Section 132(b) is a significant alternative to the usage of continuation applications as a means for further prosecution after a final rejection.

Reissue Applications

A person may apply for a *reissue* of a granted patent that is wholly or partly inoperative or invalid through an error committed without deceptive intention. While a patentee may add or amend claims through reissue, the reissue must be for the invention disclosed in the original patent and may not introduce new matter.

A reissue broadening the scope of the claims must be applied for within two years of the issue date of the original patent. A reissue that does not attempt to enlarge the scope of the claims may be applied for at any time during the life of the patent. In order to strengthen the presumption of validity, a patent owner may file a reissue application to obtain reexamination of the claims in the light of newly discovered prior art.

As an example, the validity of our claim of an illuminated hammerhead might possibly be later challenged if we failed to identify and disclose an invention that described an axe or hatchet with an illuminated head. Once we learn of this illuminated axe, we may file for a reissue that includes the disclosure of the additional prior art, and describes the advantages of our illuminated hammer over the illuminated axe. The claims may be amended as a result of the additional disclosure and amended detailed description, as well. Certain *intervening rights* may apply in the case of reissues that alter the scope of the claims.

Double Patenting

A person may not claim the same invention or obvious modifications of the same invention in more than one patent. By invalidating the second patent, the double patenting doctrine prevents a time-wise extension of the statutory period of monopoly.

However, filing a "terminal disclaimer" of the period of the second patent that extends beyond the first patent eliminates any double patenting objection. One may obtain a second patent on subject matter that is a truly patentable improvement over the invention claimed in the prior patent.

APPEALS

The Board of Patent Appeals and Interferences (BPAI) is an administrative review body within the USPTO that reviews *ex parte* appeals from adverse decisions of examiners, in those situations where a written appeal is taken by a dissatisfied patent applicant. A panel of at least three members of the BPAI hears each appeal and interference.

After an examiner twice rejects a claim in a patent application, the applicant may appeal the rejection to the BPAI, which may reverse, affirm, or enter a new ground for rejection. If the examiner makes a rejection "final," the applicant must appeal to avoid abandonment of that claim or, if all claims are rejected, of the whole application.

An applicant may seek judicial review of an adverse BPAI decision by either appeal to the Court of Appeals for the Federal Circuit or the filing of a civil review action in the District Court for the District of Columbia.

OTHER USPTO ACTIONS AND ACTS

There are a few additional important actions of which you should be aware. They are uncommon, but nevertheless may have an effect on your patent application or patent at some further date.

Interference

The USPTO may declare an "interference proceeding" when one patent application claims substantially the same patentable invention as is claimed in one or more other applications or issued patents. The purpose of an interference proceeding is to resolve the issue of priority of invention. An applicant can initiate an interference proceeding by copying the claims of a pending application or issued patent.

The Board of Patent Appeals and Interferences (BPAI) reviews interferences to determine priority (that is, decide who is the first inventor) whenever an applicant claims the same patentable invention as already claimed by another applicant or patentee.

In an interference proceeding, the burden of proof is allocated according to the order of the parties' application filing dates. Hence, there is a procedural benefit to having an earlier filing date with respect to an interference proceeding. The procedural rules followed in patent interferences are unique and relatively complex. The BPAI resolves the merits of interferences.

Interference is highly uncommon and becoming more rare each year. In fact, according to testimony presented in 2005 by the American Bar Association, the number of interferences is less than one per 1,000 issued patents. However, should this rare occurrence involve you and your patent application, contact your patent attorney.

Invention Secrecy Act

This Act prohibits anyone from filing a patent application, as to any invention made in the United States, in a foreign country prior to six months after filing an application for a patent in the United States, without obtaining a license from the Commissioner of Patents.

There are many reasons for the United States wanting the six-month window to review patent applications before allowing filing in a foreign country. For instance, given the rise in international terrorism, it would be important for the United States to review and have the opportunity to prevent the possibility of an anti-terrorism invention from falling into the hands of foreign terrorists.

In cases of inadvertent violation, the Commissioner may issue a retroactive license. The Act also authorizes the Commissioner to issue secrecy orders as to subject matter in a patent application, the disclosure of which would be detrimental to the national security of the United States.

Third Party Request for Patent Reexaminations

At the request of any person, including the patent owner, or on his or her own initiative, the Commissioner of Patents may determine that a substantial new question of patentability as to the claims of a patent has been raised by the citation of prior art patents or publications.

If the Commissioner determines that a substantial new question of patentability is raised, the claims are then reexamined according to normal examination procedures. After the reexamination, a certificate is entered canceling unpatentable claims, confirming patentable claims, and incorporating amended or new claims.

Reexamination is looked at by some as a cost-effective alternative to litigation. If a third party is successful in a reexamination bid, the Patent Office could end up invalidating your patent. You would be left with nothing.

If your patent is upheld after being reexamined, it will be very difficult for an infringer to invalidate your patent because of prior art considered by the USPTO during the reexamination proceeding. Because of this, many companies choose litigation instead of reexamination to determine the validity of a patent.

You should rely on your patent attorney to address reexamination in more detail, should this become an issue for your patent. However, at the early stage of patent drafting, you should take care to exhaustively search all prior art—both patent and non-patent literature, and make full disclosure of the prior art in your patent application. With one exception, successful reexamination depends on a third party finding prior art that you missed during your pre-filing research.

PATENT INFRINGEMENT

Under 35 U.S.C. §271(a), infringement of a patent consists of the unauthorized making, using, offering for sale, selling, or importing any patented invention within the United States. There are three types of "infringers" who may be liable for patent infringement: direct infringer; indirect infringer; and contributory infringer.

Anyone who makes, uses, or sells the patented invention without per-

JEROME LEMELSON V. CORPORATE AMERICA

It can be difficult and costly to enforce your patent against a company. However, Jerome Lemelson received over 600 patents in his lifetime and successfully enforced his patents against companies such as Ford, Dell, Boeing, General Electric, Mitsubishi, and Motorola, receiving over $1.5 billion in licensing fees.

mission is a *direct infringer.* Anyone who "actively induces infringement of a patent" by another is an *indirect infringer.* Anyone who sells a product knowing the same to be especially made or adapted for use in an infringement of a patent is a *contributory infringer.*

As stated previously, the claims of a patent define the scope of coverage against infringers—similar to the metes and bounds of a land description. The broadest valid claim(s) is typically utilized to determine infringement of the patent. The accused infringing product must "read on" every element of the broadest valid claim in the patent.

When finding infringement on a utility patent, the meaning of the claims of the patent is determined by the trial judge. When the meaning of a claim is contested, many courts will hold a preliminary hearing to resolve, as a matter of law, the meaning of the claim which defines the scope of the patent grant. This usually requires expert testimony, either technical or patent or both.

When only insignificant differences exist between the elements of the claim and the infringing product, the Doctrine of Equivalents may be utilized to find infringement. The purpose of the doctrine is to catch unscrupulous copiers who make unimportant and insubstantial changes and substitutions which, though adding nothing, avoid the literal language of the claims. Finding infringement on a design patent requires determining if an ordinary observer would be deceived by two substantially similar designs.

It is important to perform patent searches for patents that will prevent the patentability of, and raise potential infringement issues for, your invention. In addition, you should review competing products and marketing materials for patent notices (where they expressly state the patent numbers that protect the products). Your patent attorney can assist you in determining if there are any potential infringement issues prior to marketing your product.

We've covered the scope of the important legal processes, timelines, and procedures that you must follow to ensure that you file properly, and on time. If your application fails to meet the governing requirements, or if you miss any important deadlines, the validity of your patent could be at risk. After learning how to write a complete and powerful patent application, the last thing you would want to have happen is for your application to be denied because of a legal or procedural technicality.

Refer back to this chapter throughout your patent-writing process to make sure you are complying with these provisions, and as always, periodically meet with your patent attorney to review any substantive changes to patent laws of which you need to be made aware.

Conclusion

As you have discovered, there's a lot to learn about drafting a successful patent application. It may even seem insurmountable at times. However, once you overcome the difficulties that arise and complete the process, it is an exciting accomplishment. By reading *The Patent Writer* you have taken a step in the right direction. Your commitment to seeing it through will also make each subsequent patent application a lot easier to draft.

You'll find, like most experienced inventors and engineers, that you learn from your mistakes and the challenges of drafting an application with a broad scope. If you feel overwhelmed, just remember:

- Don't despair.

- Take some time to rethink your approach and alternative ways of describing your invention.

- It gets a lot easier as time goes on.

Should you have any questions regarding this book, we encourage you to contact us through any of our websites:

> **www.frompatenttoprofit.com** (IPT Company)
>
> **www.patentcafe.com** (PatentCafe.com, Inc.)
>
> **www.neustel.com** (Neustel Law Offices, LTD)

In closing, we sincerely wish you the very best in your invention and patent-drafting activities!

Glossary

Abandonment. A forfeiture of an application or invention with regards to the governing patent office. This may be accomplished by some positive act or failure to act within a reasonable or statutorily fixed time and may be either expressed or implied.

Absolute Novelty. A patent requirement wherein prior public disclosure or sale anywhere in the world before filing of a patent application within most developed countries, except the United States, will be a bar to obtaining a valid and enforceable patent in those countries.

Abstract. A brief summary of an invention in the body of a patent application that will quickly identify its key features. It is used to define the general scope and operation of the invention.

Amendment. A response by the inventor or patent attorney to an office action by a United States Patent and Trademark Office examiner, usually making necessary changes to either the text of the application or to the drawings.

Application for Patent. *See* Patent Application.

Assignment. Occurs when you sell or bequeath your intellectual property rights to someone else, such as an employee who assigns patent rights to an employer. This is simply a change of ownership of property.

Assignor. One who assigns or transfers a patent right to another person or entity. This is accomplished through an Assignment.

Background of the Invention. The section in a patent application that describes the general nature of the invention. Typically it is broken down into two subsections entitled "Field of the Invention" and "Description of the Related Art."

Best Mode. A condition for the issue of a patent. An inventor must describe the best mode he or she knows for carrying out the invention. If the inventor discovers, but does not disclose, a better method of implementing the invention prior to filing the application, any resulting patent could be invalidated.

Block Diagram. A type of drawing used in patent applications that shows the interrelationship between the components used in an invention.

Blown-Up View. A drawing type that has been substantially enlarged in order to view the inventive matter sufficiently.

Board of Appeals. An administrative board of senior patent personnel that hears appeals from applicants and reviews the decisions of examiners on applications for patents. If an applicant or his or her representative disagrees with the ruling of examination, the Board of Appeals may review the application in question.

Brief Description of the Drawings. A section heading in a patent application listing the drawings in order, describing the details shown in the drawings.

Brief Summary of the Invention. The section of a patent application describing an invention in broad terms, such as how it is composed and how it works.

Broad Claim. A patent term used to describe the scope of a patent claim. A claim is identified as a broad claim when it incorporates a large range of alternative embodiments of the invention with the provision that those embodiments are described within the specification.

Business Methods. A patent on a method of conducting business. May also include training systems and methods, including those encompassing computer and software methods.

CFR. The United States Code of Federal Regulations.

CIP. *See* Continuation-In-Part.

Citation. A reference to patents or journal articles that the applicant or examiner deems relevant to a current application. References to legal authorities or prior art documentation are examples of a citation.

Claim. A written statement at the close of a patent application specifically stating what the inventor alleges as the invention. The claims define the legal scope of a patent. What falls within that definition is protected by the patent. Anything outside the scope of the claim is not protected.

Claim Statement. A statement prepared before writing a patent application outlining the claims the inventor will be pursuing. Sometimes referred to as a "claim plan."

Commercialization. A process of taking an invention to the marketplace with the intention of deriving revenue. This will consist of protection, design, development, and marketing, then establishing the product in a viable distribution system to distribute the finished product.

Complete Specification. *See* Specification.

Composition of Matter. Usually scientific by nature, such as plastics and bio-engineering. However, they may include blended compositions used to give a particular product certain performance characteristics.

Continuation-in-Part (CIP). A child application that claims priority from a parent application, but has additional subject matter not disclosed in the parent application. This form of application may be used to add improvements that were developed after the filing of the parent application.

Continued Prosecution Application (CPA). A continuation or divisional application filed under 37 CFR 1.53(d) that continues the prosecution of the original patent application. A continued prosecution application cannot contain additional subject matter.

Copyright. Covers any written material, designs, patterns, songs, video, and sculptures. Copyrights are filed with the U.S. Copyright Office. Rights are usually established the moment they are created.

CPA. *See* Continued Prosecution Application.

Date of Original Conception. The date in which an inventor can legally verify and prove the conception of an invention.

Declaration. A declaration is included in a U.S. patent application stating the inventor's belief that he or she is the first and original inventor. This must be done for each inventor listed on application. *See also* Oath.

Dependent Claim. A claim in a patent that refers to a prior claim and inherently includes the limitations of the prior claim. It defines an invention that is narrower in scope than in the previous claim.

209

Description of Related Art. A subsection of the "Background of the Invention" and includes descriptions of pertinent art related to the invention disclosed in the application.

Design-Around. The term used to describe products that do not include the patentable subject matter disclosed in a patent. Generally speaking, the broader the claims in a patent, the more difficult it is to design-around.

Design Patent. A type of patent that covers only the ornamental aspects of an item. A design patent may be granted for any new, ornamental, and original invention. Any functional aspects of an invention must be claimed by a separate utility patent. Both design and utility patents may be obtained on an invention.

Detailed Description of the Invention. The text section of a patent application that includes a complete description of the best mode of the invention. This is a very important part of a patent application, as additional subject matter (new matter) cannot be added to the application once it is filed.

Device Patents. A patent covering the physical form of an invention that defines the structure, apparatus, or composition of a product. Same as product, apparatus, or structure patent.

Diligence. After recording the date of original conception, an inventor must show diligence as it is reduced to practice. Without sufficient diligence, a claim of abandonment may result, thus the loss of the inventor's conception date as the priority date.

Disclosure. A public dissemination of information about an invention, by publication, usage, or other mode. An adequate disclosure should enable reproduction of the device or process to actual reduction of practice through only the information disclosed.

Disclosure Document Program (DDP). A USPTO service that allows inventors to record information about their invention for up to two years, with the purpose of establishing dates of conception, reduction to practice, progress, improvements, etc. This is not a patent application and does not provide patent protection.

Divisional Application. Only one invention may be claimed in a patent application. The examiner may impose a restriction requirement in which the applicant must chose which one invention will continue as the current application, following which the applicant may file a second patent application, called a divisional application.

Doctrine of Equivalents. Under this doctrine, a potential infringer cannot circumvent patent claim by the mere usage of a similar process that essentially does the same thing as the invention claimed.

Dominant Patent. Refers to a patent that has a broader scope than other related, narrower patents that may be considered improvements. Usually it is a previously issued patent with a broader scope.

Double Patenting. An attempt to obtain a second patent on the same invention. Double patenting is found when the claims of both patents, by their descriptions, cover essentially the same things. This can typically be overcome by filing a Terminal Disclaimer.

Drawings. Figures, illustrations, or visual diagrams of an invention that are included in a patent application for the purpose of conveying a more complete understanding of the device or method. Patent drawings must be included in the application, unless the nature of the invention precludes them. Special rules and techniques must be followed in the preparation of patent drawings.

Effective Date. That date when an application or patent takes effect and becomes enforceable, or the date by which a prior art reference (a publication or patent) can be applied against an application to act as a bar to the allowance or patentability of the application's claims.

Election. When an examiner makes a restriction requirement because more than one invention is being claimed in a patent application, the applicant must choose, or elect, one invention to continue prosecution of the patent application. A divisional or continuation application may be filed for the non-elected invention(s).

Enablement. An invention must be described in the specification of a patent or patent application with sufficient detail to enable someone with ordinary skill in that art to reproduce and utilize the invention without undue experimentation.

EPO. The European Patent Organization (sometimes referred to as the European Patent Office).

Examination. The study of a patent application in the governing patent and trademark office, by a patent examiner, to determine whether the invention described therein can be patented. The major considerations the examiner addresses are the novelty and utility of the invention, among other things.

Examiner's Amendment. At the examiner's prerogative, minor errors in an application may be corrected. This usually happens only after the examiner deems the application patentable.

Exploded View (Drawings). A drawing that illustrates the interrelationship between two or more components, such as assembly drawings.

Family of Claims. The various patentable subject matters described and claimed in a patent application.

Fee. Various fees are required during the life of a patent. Some of these fees include filing fee, issuance fee, maintenance fee, and petition fee. Fees differ according to the country that governs.

Field of the Invention. The definition of the specific field or art to which an invention pertains. *See also* Background of the Invention.

Fig. *See* Figure (Drawings).

Figure (Drawings). The proper designating term to use when referencing drawings in a patent application, often shortened to Fig. For instance, "Fig. 1 is a perspective view."

File Wrapper. The term applied to the "entire folder" containing all correspondence between the examiner and inventor (or the appointed attorney), as well as any models or other materials that may have been requested. *See also* Prosecution History.

Filing Date. The date when a properly prepared application is officially filed with the patent office. The filing date is typically the day the patent application is mailed to the USPTO if U.S. Express Mail is utilized along with a Certificate of Express Mail. Under U.S. patent laws, this is the beginning of the twenty-year term of patent protection.

Filing Receipt. A document sent by the USPTO, usually within three months of filing a patent application, that serves as evidence of the filing of a patent application. The filing receipt identifies the application serial number, filing date, invention title, and inventor(s). It also typically grants a Foreign Filing License to the inventor, thereby providing a required license to the inventor to file for patent protection in foreign countries.

Final Rejection. The decision by a patent examiner to terminate further prosecution of an application. An applicant can only amend the application to place the claims in condition for allowance. An applicant also has a right to an appeal.

First-to-File. The filing date of a patent application, and not the date of invention, determines when an inventor's rights begin. Most countries other than the United States and the Philippines have first-to-file patent systems.

First-to-Invent. The U.S. and Philippine patent systems recognize an inventor's right to a patent based upon the date of actual invention, as opposed to the first-to-file system used by most other countries.

Flow Chart (Drawings). A type of drawing comprised of a series of shapes used to define various operations and functions. Utilized to illustrate the flow of activity such as in a manufacturing, business method, electrical, or software process.

Foreign Filing License. A license granted by the USPTO that allows the inventor to file for a foreign patent. The foreign filing license is required before an inventor can file a foreign patent application. If a U.S. patent application or PCT application is filed with the USPTO, the foreign filing license is usually granted or denied in the Filing Receipt.

Formal Drawings. Patent drawings are the perfected ones made exactly according to U.S. Patent Office rules. Formal drawings are not required to be submitted with the initial patent application. *See also* Drawings.

Front View (Drawings). The drawing type used to illustrate the one-dimensional frontal view of an invention.

Functionality. The design aspect that requires that a product must work better for its intended purpose, as opposed to cosmetic improvements to the product.

Generic Claim. Type of claim that relates to a whole group or class. A claim to a generic invention may include within its scope the subject matter of narrower claims.

Harmonization. Through one worldwide system of patent laws, all patent laws of all countries of the world would be adjusted or modified to produce harmony of the world patent system. This is only a proposed concept at this point.

Independent Claim. A patent claim that has no dependency upon another claim within an application.

Information Disclosure Statement. A document disclosing prior art to the USPTO that, during the pendency of an application, is "material" to the patentability of the invention. It must be filed within three months of the patent application filing date, or before the first office action with the exception of paying a fee.

Infringement. An invasion of an exclusive right of intellectual property. An unauthorized use of the invention described in a claim of a valid patent without proper license or consent of the owner of the patent rights.

Injunction. A court order prohibiting someone from a specified act, such as infringing upon a patent.

Insufficient Disclosure. A condition in which an application or patent contains an incomplete description of an invention in its specifications. A disclosure is deemed insufficient when a person skilled in the art is unable to reduce the invention to practice with information and material disclosed in the application itself.

Intellectual Property. Creative ideas and expressions of the human mind that have commercial value and receive the legal protection of a property right. Types of intellectual property include patents, trademarks, designs, confidential information, trade secrets, copyright, circuit layout rights, plant breeder's rights, etc.

Invention. The creation of a new technical idea and of the physical means or process to accomplish or embody it. An invention described in a patent application must contain, at a minimum, every element found in the patent claim.

Invention Disclosure. The written disclosure of a novel, useful, unique invention. The first disclosure usually qualifies the date of original conception.

Inventive Matter. Commonly referred to as the "novel, useful, unobvious" subject matter that describes an invention in a patent application. More specifically, it may be called the "patentable subject matter," should it be considered patentable.

Inventor. Anyone whose involvement and contribution was essential to the development of the invention. The one who is first to conceive of a particular invention and who diligently works to convert this conception into a tangible physical property.

IP. *See* Intellectual Property.

ISA. The "International Searching Authority" for the PCT is either a national patent office or an intergovernmental organization such as the EPO.

ISR. "International Search Report," a published report following a patentability search completed for a PCT, which is later used by the national patent office that will issue the patent.

Issue Date. Not to be confused with the filing date, this is the date on which the patent actually issues.

214

Joint Inventor. Two or more inventors of a single invention who work together in the inventive process. Mere assistance in developing an invention does not make one a joint inventor. Inventorship is defined by the individual(s) that contribute to the subject matter identified in the claims.

Knock-Off. A copy of a work or product that is protected by patent, trademark, trade dress, or copyright, which is indistinguishably similar.

Large Entity. Companies with over 500 employees. They pay the full amount for all patent related fees.

Lead Lines (Drawings). The drawn lines that reference a particular attribute or element in a drawing in a patent application. The lead line should touch the specific attribute or element. All lead lines have a corresponding number, which is referenced in the body of the patent application.

Lead Lines with Arrows (Drawings). A lead line with an arrow that is floating refers to the entire object (invention) in a drawing. A lead line with an arrow that touches a certain part of a drawing refers to the "entire component." All lead lines with arrows have a corresponding number, which is referenced in the body of the patent application.

Legal Monopoly. The term commonly given to the rights endowed by a U.S. patent. Thus, you can say the owner is given the right to exclude others from manufacturing, using, and selling products based on the scope of the patent protection. Also commonly referred to as "negative rights."

Lettering (Drawings). Commonly used in patent drawings to signify a force or physical property related to the invention. Also commonly used in prior art drawings in an application to describe the various embodiments.

License. A contractual agreement giving written permission to another party the right to use an invention, creative work, or trademark. By licensing an invention or work to a company, you get money (often in the form of royalties) in return for allowing the company to use, produce, and sell copies of your invention or work in the specified marketplace.

Machine Patent. A patent granted on an invention considered a machine. Machines typically consist of a group of various components that produce a desired outcome. Frequently accompanied by process patents.

Maintenance Fees. The patent fees needed to keep a utility or plant patent in force for its full life. Different patent offices charge different fees and have different intervals of payment. U.S. design patents do not have maintenance fees.

215

Manual of Patent Examination and Procedure (MPEP). The USPTO manual is utilized by examiners to review and examine patent applications. The current MPEP may be viewed at www.patentwriter.com.

Mask Works. Circuit layout rights that automatically protect original layout designs for integrated circuits and computer chips. These are a separate form of protection from conventional patent, trademark, and copyright laws.

Mathematical Formula. Patentable subject matter based on a certain unique mathematical sequence that produces a desired result. Commonly used in software and business methods patents.

Means. A term commonly used when writing patent applications in order to expand the breadth of scope of the inventive subject matter. For instance, instead of saying, "Part A is affixed to Part B by glue or adhesive or nails or rivets . . . ," you may say, "Part A is affixed to Part B by a 'means of adjoinment' such as glue, rivets, and so on."

Method of Use. Also commonly referred to as a "systems patent." This illustrates a new, unique use of an invention. May consist of one or more prior art components.

MPEP. *See* Manual of Patent Examination and Procedure.

Narrow Claim. When the scope of a claim is specifically defined, it is referred to as being "narrow in scope." Typically narrow claims are easier to design-around.

National Phase. The national phase of the PCT process is where you have to file initial patent application files in specific countries. It should be noted that there is no "world-wide patent" and that you must prosecute a patent application in each country that you want patent protection in, even through the PCT process.

Negative Limitation. The description of an invention describing what it will not do or does not contain, as opposed to a description of what the invention will do or contain.

Negative Rules of Patentability. These rules say a patent cannot be granted merely for such concepts as substitution of materials, performing automatically what had been done manually, rearrangement of parts, or a change in proportions.

New. *See* Novelty.

New Matter. Usually used with regards to an amendment to a pending application in order to change the body of application. No additional material may be added by amendment to a patent application after filing if it adds, deletes, or changes the description of the invention, its embodiment, or best mode of operation. However, amendments that clarify or further define originally disclosed material are acceptable and are not new matter.

New Property or Use. A patent cannot be obtained for a new property or use of a previously known composition of matter.

Non-Elected Invention. An invention that is not selected for continued examination due to a restriction.

Non-Essential Material. Material incorporated into an application not essential to the art being reduced to actual practice or having no bearing on the validity of the application itself.

Non-Obvious (also referred to as Unobvious). A patent is not permitted if the subject matter for which the patent is being sought would have been obvious to an ordinary person skilled in the art, as a whole, at the time the invention was made. This would include someone attempting to patent an application in one field that had previously been obvious in an unrelated field.

Non-Publication Request. A request by an applicant to the USPTO that his or her patent application not be automatically published after eighteen months of pendency. This request must be included with the patent application and cannot be filed at a later date. Without filing a Non-Publication Request, a patent application will be automatically published by the USPTO unless the applicant expressly abandons the application.

Not Anticipated. Relating to an invention claim. It is said that it is novel and passes the requirements of 35 USC 102 if the invention is not anticipated by prior art.

Not Shown. A term describing a certain component or element in a patent drawing that may not be visible (or is hidden) from the drawing's perspective. Thus, the writer refers to the components as "not shown." When using this term, the component or element would be described in some other drawing, in another view, if the referenced component is essential to the invention.

Notice of Allowance. When a patent office officially notifies a patent applicant that an application's claims have been allowed and a patent will be granted. Normally an issuance fee must be paid within a preset time for the patent actually to issue.

Novelty. A requirement for patentability. A claimed invention is novel and passes the requirements of 35 USC 102 if the invention is not anticipated by prior art. If a claim is anticipated by any single reference in the prior art, it then lacks novelty. Many patent offices require that for a claim to have novelty, it must not have been revealed or publicly available anywhere in the world.

Oath. A declaration included in a U.S. patent application stating the inventor's belief that he or she is the first and original inventor that is notarized. This must be done for each inventor listed on application. Also commonly referred to as "Oath of Inventorship." *See also* Declaration.

Objections. When an examiner takes exception to any matter in an application where he or she may require correction. An application that is stalled in prosecution usually is because of an examiner's objection.

Obviousness. A condition in which an invention cannot receive a patent because a person with ordinary skill in the art could easily formulate it from publicly available information, such as prior art.

Office Action. Official action on a patent application is a written communication (office action) from a patent office.

On Sale Bar. A statutory bar in which an inventor cannot obtain a patent if he or she waits for more than one year to file a patent application after a product embodying the invention has been placed "on sale." This would include any public disclosure of the invention, whether the product is offered for sale or not. Also commonly referred to as the "one year rule." There are exceptions to this rule, such as in situations where the offer for sale or sale was for the purposes of "experimentation."

Opposition. A third party may request an application be refused or a granted patent be revoked. This may be based on the grounds of discovery of impropriety of originality, novelty, or prior public disclosure.

Ordinary Skill in the Art. An engineer, scientist, or designer in a technology that is relevant to an invention that possesses an ordinary skill or level of technical knowledge, experience, and expertise. This is used as a benchmark in evaluating skill level as it relates to intellectual property development.

Ornamental Appearance. Refers to a design patent that covers only the ornamental appearance of an invention, and not its utility.

Parent Application. A primary patent application is referred to as a "parent" application. This is in order to differentiate it from a continuation or continuation-in-part application.

Patent. A legal monopoly, granted by a county's Patent and Trademark Office (PTO), for the use, manufacture, and sale of an invention for a specific period of time. At the end of the term (in the U.S. it is twenty years from filing date for utility patents and fourteen years from the date of issuance for design patents), the technology becomes public property. Patents do not protect ideas, only structures and methods that apply technological concepts. There are three kinds of patents in the United States: a utility patent on the functional aspects of products and processes; a design patent on the ornamental design of useful objects; and a plant patent on a new variety of a living plant.

Patent Agent. A person with technical training and experience, and who demonstrates an understanding of patent law, but who is not an attorney. They must pass a requisite examination and must be certified to practice patent law in order to represent others before a patent office. Patent agents cannot appeal examiner decisions to the Board of Appeals, provide legal opinions, or initiate legal actions.

Patent Application. An application for the protection of an invention; those documents or papers including a petition, a specification, drawings (when required), one or more claims, an oath or declaration, and requisite filing fee, with which an applicant seeks a patent.

Patent Attorney. An attorney at law who has also met the requirements of training, experience, and passing an examination, and been certified as a patent agent; a patent attorney may represent a client in civil, business, and other interactions (such as licensing, sale of a patent, or other actions between individuals) in addition to those between an inventor and the USPTO.

Patent Claim. *See* Claim.

Patent Cooperation Treaty (PCT). A united agreement between many countries. The PCT was created in 1978 to allow an inventor to file one international patent application in one member country while designating the application for one or more member countries. The benefit of this Treaty is a simplified process for obtaining international patents, reduced work, and elimination of duplicate efforts on the part of member countries and the applicant.

Patent Drawing. *See* Drawings.

Patent Family. The basic patent and all equivalent patents for the same invention in more than one country. The family relationship need not be ascertained through priority information in order to be included as part of a patent family.

Patent Law. Title 35 of the United States Code.

Patent Pending. A mark applied to any product to let the purchaser know that a patent application has been applied for some portion of that article or some process related to that article. Some countries do not permit the use of this form of mark. No actual patent protection is in force during pending status.

Patent Protection. Based on the scope of the claims in a patent, protection is provided to the owner of a patent. The scope defines the degree of protection and the ability of the owner to exclude others from making, using, and selling the invention.

Patent Search. *See* Search.

Patent and Trademark Office (PTO or USPTO). *See* United States Patent and Trademark Office.

Patentability. An examination of the publications and patents in the U.S. Patent and Trademark Office to determine the probable patentability of the invention. The three basic requirements for patentability for utility patents are utility, novelty, and non-obviousness.

Patentable Subject Matter. The U.S. allows patents in four primary categories of subject matter: machines, manufactures, compositions of matter, and processes; the first three categories address structural subject matter while the last covers operational subject matter. Abstract ideas, laws of nature, and natural phenomena are not patentable subject matter.

Patentee. The inventor or one who has received rights to the patent (assignee). For the applicant to be considered a patentee, the patent must have been issued.

PCT. *See* Patent Cooperation Treaty.

Pending Application. An application filed with a patent office that is pending issuance. The application is actively being prosecuted or is in the process of an appeal.

Perspective View (Drawings). A three-dimensional drawing type used to illustrate an invention. May also be referred to as "side perspective, front perspective, rear perspective, upper perspective," and so on. The most commonly used drawing type.

Phantom (Drawings). The term used to describe a component in a drawing in its invisible state by the use of dashed lines.

220

Plan View (Drawings). The drawing type used in a patent's drawings of an invention taken from a flat view above. Typically this would be a view of an object that is rather flat. *See also* Top View.

Plant Patent. A patent issued for new strains of asexually reproducing plants. Tuber propagated plants or uncultivated (wild) plants may not be patented. Some countries may not allow plant patents.

PPA. *See* Provisional Patent Application.

Preferred Embodiment. *See* Best Mode.

Present Invention. A term referring to the invention that is being revealed and described in a patent application.

Prior Art. The total body of knowledge that exists prior to the conception of a new invention. Prior art includes documentary sources such as patents and publications from anywhere in the world, and non-documentary sources such as things known or used publicly, and is referenced to determine patentability of a new invention.

Priority Date. Normally the priority date is the earliest date a patent application is filed. The priority date of a permanent patent that read on an earlier provisional patent application would be the date of filing of the provisional patent application.

Pro Se. An inventor who files an application without the assistance of a qualified patent attorney files "pro se."

Process Patent (or Claim). A patent that covers the process in which an invention is made or the process in which it performs, in contrast to a product or an apparatus patent, or method of use or systems patent. For example, computer software may claim the actual uniqueness of its formulae and the process it performs.

Product Patent (or Claim). A patent covering the physical form of an invention that defines the structure, apparatus, or composition of a product, in contrast to a process or a method. Same as device, apparatus, or structure patent.

Prosecution. This sums up the entirety of proceedings and action by the patent attorney and the internal action of the Patent Office that is reviewing a patent application. Prosecution starts when the Patent Office receives the application, and ends when the Patent Office either issues the patent, or terminates prosecution through examination or appeal.

Prosecution History. Comprises all documents relating to the application for and review of a patent, including other documents and letters generated with regard to the application.

Provisional Patent Application (PPA). An interim patent application that provides a one-year period of "patent pending" for product development. A PPA does not require claims or a declaration. A PPA has a significantly lower filing fee than a permanent patent application. It provides the legal effect of an early patent application filing date for an invention. The PPA does not take the place of a regular patent application, but it does confer patent pending status on the underlying invention.

PTO. *See* United States Patent and Trademark Office.

Public Domain. When an invention, creative work, commercial symbol, or any other creation is not protected by some form of intellectual property laws. These items become public domain and are available for copying and use by anyone.

Public Knowledge. Used in determining novelty. A patent is barred if an invention was previously known by anyone that is skilled in the art that pertains to the specific invention. This may be through prior art or publication.

Public Use. An official bar that prohibits the issuance of a patent if anyone uses an invention in public more than one year prior to filing a U.S. patent application or in the case of use of an invention for profit, where the use is not purposely hidden. Most countries will not award a patent if there is any public use of the invention prior to the application date.

Publication. The distribution or disclosure in a form that is readily accessible or distributed to the public of copies, audio recordings, or any creative work.

Publication Date (Patent Application). The date that a patent application is published by a patent office. Patent applications are published in most foreign patent offices after a specified period of time. Patent applications in the United States are published after eighteen months unless the applicant requests Non-Publication at the filing of the patent application. *See also* Non-Publication Request.

Reads On. For a product to infringe upon a claim, the product must have elements that "read on" all of the elements of the claim. In other words, if Claim 1 has elements A, B, C, and D, an infringing product will also have elements A, B, C, and D; it therefore "reads on" the claim.

Rear View (Drawings). The view of the rear, or the back side, of an object.

Reduction to Practice. There are two types of reduction to practice: actual reduction to practice and constructive reduction to practice. Actual reduction to practice is the actual implementation of the technology or process.

Constructive reduction to practice is the filing of a patent application regardless if the invention has been physically created or practiced.

Reference. A document or patent that an examiner cites against an application's claim.

Rejection. An office action by the patent examiner stating to the applicant or applicant's attorney that a claim in the patent application does not comply with the requirements for patentability.

Restriction. An office action in which an examiner may impose a restriction requirement if an application claims more than one independent and distinct invention. *See also* Election.

Reverse Engineering. This process is used to determine what something is made of, what makes it work, and how it was produced; it is accomplished by reversing the normal steps of engineering. If done properly, this method is entirely legitimate and legal with respect to inventions protected solely by trade secrets, but not for inventions protected by patents.

Scope. Refers to the breadth of a claim in a patent. *See also* Broad Claim and Narrow Claim.

Search. A study of patent materials to locate patents and other literature that pertains to an invention for the purpose of determining if any prior discovery makes the subject invention incapable of being patented or, if patentable, whether it infringes a prior issued patent. Among the different types of search categories would be a state of art search, a patentability search, an infringement search, and a validity search.

Serial Number. An identification number assigned by the Patent Office to an application on the date it is received or made complete.

Service Mark. The same as a trademark, except utilized to identify the source of services. *See also* Trademark.

Shading (Drawings). Shading is utilized within patent drawings to indicate surfaces, material types, shapes, and contours.

Side View (Drawings). The one-dimensional view of the side of an object. Sometimes referred to as an "end view."

Similar and Same Elements (Drawings). In a patent drawing, similar and same elements may either have the same number as an element in a previous drawing or a corresponding multiple of ten. Same elements in a single drawing, such as multiples of a tooth in a gear will be described with the same number, but with a small case letter attached, such as 38a, 38b, 38c, and so on.

Small Entity Status. The PTO charges small entities (a for-profit company with 500 or fewer employees, including independent inventors) half the fees charged large entities for filing a patent application and for issuing and maintaining the patent.

Specification. The written description of an invention describing the invention in sufficient detail that another person could reduce it to actual practice. The specification must describe "the best mode contemplated by the inventor for carrying out his invention." Drawings are included in the specification when required.

State of the Art. Comprises all available knowledge and understanding that exists in a given field of technology at the time an invention was made.

Subject Matter. *See* Patentable Subject Matter.

Summary of the Invention. *See* Brief Summary of the Invention.

Systems Patent. *See* Method of Use.

Title of the Invention. The title of an invention in a patent application must be descriptive and not a trademark or nickname.

Top View (Drawings). The one-dimensional view of on object looking at it from the top. *See also* Plan View.

Trade Secret. Any formula, pattern, machine, or process of manufacturing, or any device or compilation of information used in one's business, which is maintained in secrecy and which may give the owner an advantage over competitors who do not know or use it. Such confidential information is protected against those who gain access to it through improper methods or by a breach of confidence. A legitimate way of bypassing a trade secret is through reverse engineering.

Trademark. Any identifying symbol, including a word, design, or shape of a product or container, that qualifies for legal status as a trademark, service mark, collective mark, certification mark, trade name, or trade dress. Trademarks identify the source of goods or services. Trademarks distinguish one seller's goods from goods sold by others.

Unique. All inventions must be considered unique in order to qualify for patent protection. Inventive matter is not considered unique if it is obvious to those skilled in the art, or is considered as being anticipated by previous patents or existing products.

United States Patent and Trademark Office (PTO or USPTO). The office charged with overseeing and implementing the federal patent and trademark laws. Its principal responsibility is to examine applications for patents and the registration of rights in trademarks, service marks, certification marks, and collective marks. Also it issues all patents and trademarks in the United States. The U.S. Department of Commerce originally established the PTO in 1836.

USC. The United States Code.

Useful (or Usefulness). A utility patent is required to have some purpose. In the case of design patents, usefulness may be ornamental. The useful purpose can be solely for amusement or a minor improvement on an existing design.

USPTO. *See* United States Patent and Trademark Office.

Utility. Defines usefulness of a patented invention. An absolute requirement in determining patentability for a utility patent application. This is not required for receiving a design patent.

Utility Patent. The most commonly issued type of patent. A utility patent may be granted for any new, useful, and non-obvious invention. According to 35 USC 101, utility patents are subdivided into mechanical, electrical, and chemical categories.

World Intellectual Property Organization (WIPO). A Union of member countries formed by the Paris Convention for the Protection of Industrial Property. WIPO, located in Geneva, Switzerland, is responsible for promoting the protection of intellectual property throughout the world.

Resource List

The following is a list of useful resources for the inventor. They are arranged by category.

AUTHORS' WEB RESOURCES

The Patent Writer™
Visit the official website for this book for downloadable forms and important U.S. Patent and Trademark Office updates related to filing patent or provisional patent applications. See www.patentwriter.com

From Patent To Profit
Publishes the *Scientific Journal,* an approved inventor's journal that guides inventors through the first important steps of documenting their new invention. Visit www.frompatenttoprofit.com

PatentCafe®
PatentCafe's *ICO Global Patent Search* is a powerful fee-based commercial patent search engine used by professional inventors, patent attorneys, and Fortune 500 companies for prior art searching. Using advanced linguistics search engine technology, the search engine automatically generates a set of related thesaurus words for each patent that appears in the search results, and analyzes search results by creating charts and graphs for viewing companies practicing in your invention field. Visit www.PatentCafe.com

IPFrontline™ Free Online Magazine

With over 6,000 inventor help articles available online, *IPFrontline* is the largest web-based collection of invention-related articles available anywhere. Gain an understanding of patent licensing, prototyping, patenting, patent searching, and patent strategy—from the perspective of other inventors as well as that of the big corporations to which you hope to license your invention. An annual subscription is required for accessing the full article archive or industry patent analytics. Visit www.IPFrontline.com

Neustel Law Offices, LTD

Michael Neustel is a U.S. Registered Patent Attorney who assists inventors and businesses in patenting their invention. Visit www.neustel.com

PatentHunter

PatentHunter™ is a software program that helps patent businesses and inventors search, download, and manage United States and foreign patents. You may download a trial version, or purchase an annual subscription. Visit www.patenthunter.com

PatentWizard

PatentWizard™ software assists individuals in the preparation and filing of provisional patent applications by leading the inventor through the sequential path similar to what we describe. Visit www.PatentWizard.com

Patents in Commerce™ (PIC)

PIC is a complete invention development, patent, and intellectual property training series consisting of DVDs, Research Workbooks, invention and patent books, and other materials and software. The authors and presenters are a team of highly experienced inventors, marketers, attorneys, prototypers, and government specialists who show inventors how to successfully commercialize their innovations. Experts present a process-driven approach that shows the various aspects of product development and design, patenting, prototyping, manufacturing, invention financing, licensing, and product marketing, which are needed to successfully commercialize an invention. Visit www.patentsincommerce.com

PATENT SEARCH TOOLS

United States Patent and Trademark Office

Provides free online patent searching, plus in-depth resources for inventors, including comprehensive information about patents and trademarks, frequently asked questions, and helpful brochures. Visit www.uspto.gov or call 800-PTO-9199.

European Patent Office

Search more than 50 million patents through the European Patent Office's free online patent search engine. Visit www.espacenet.com/

Internet search engines are valuable non-patent literature resources in determining the patentability of your invention. Try these popular ones:

- www.google.com
- www.alltheweb.com
- www.yahoo.com
- www.hoovers.com

Search catalogs, stores, and other materials where products similar to your invention may have been promoted.

USING A THESAURUS

A thesaurus can come in handy when first conducting your preliminary patent search. A thesaurus can be a valuable tool in determining the proper words and phrases you should utilize in your patent application. Use thesaurus words to expand your vocabulary in an effort to broaden the description and claims of your own patent.

Many word-processing programs have a built-in thesaurus you can use (though it is limited). You can also utilize free Internet-based thesauruses such as www.thesaurus.com.

USING ACRONYMS

You can greatly simplify your patent application by assigning initials to word groups that describe various parts of your invention. It is sometimes easier to write and understand a patent application that efficiently utilizes acronyms.

You can utilize acronyms that are widely recognized in your industry (e.g. HVAC—Heating, Ventilating, and Air Conditioning). You can also create your own acronyms (e.g. RCFM—Remote Controlled Flying Machine).

The number and types of acronyms you utilize is unlimited. However, you need to identify the particular acronym you plan to utilize within the application so there is no confusion, as some acronyms have more than one meaning. Be careful that you do not create an acronym that is widely used in an industry to describe something entirely different from your own intended use. You can quickly check to see if your acronym falls into this category by conducting an acronym search at www.acronymfinder.com.

INVENTOR'S JOURNAL

The United Inventor's Association offers a simplified inventor's journal available through their bookstore. As the Association is a highly trusted organization, the UIA/USA's inventor's journal provides all of the necessary features to fulfill the Patent Office's requirements, but does not include the advice and guidance of the Scientific Journal listed above. Visit www.inventorhelp.com/store/index.html

PATENT ATTORNEYS AND AGENTS

USPTO's List of Attorneys and Agents Registered to Practice Before the USPTO is a publication provided by the U.S. Patent Office and is available at many Patent and Trademark Depository Libraries. Call the U.S. Patent Office at 1-800-PTO-9199 or visit www.uspto.gov/main/profiles/patty.htm

FLOW CHARTS

For a more in-depth understanding of the various forms of boxes, symbols, and line types used in flow charts, visit the Deming Quality Management website at http://deming.eng.clemson.edu/pub/tutorials/qctools/flowm.htm#Overview

GOVERNMENT ASSISTANCE AND INVENTOR ORGANIZATIONS

United States Patent and Trademark Office (USPTO)
The USPTO has developed a new website dedicated to helping inventors and small businesses better identify and address their intellectual property protection needs. The USPTO is also working with organizations like the U.S. Chamber of Commerce and the National Association of Manufacturers to help spread the word about the benefits of filing for IP protection.

Free informational materials, which can be downloaded from this website, will help guide small businesses through the often-complicated world of intellectual property protection. Visit www.uspto.gov/smallbusiness/

Small Business Development Centers (SBDC)
SBDCs are almost always willing to provide strong counsel to inventors and innovators. There are more than 1,000 offices throughout the United States. Visit www.sba.gov/gopher/Local-Information/Small-Business-Development-Centers/

The United Inventor's Association of the USA
UIA/USA was formed as a national umbrella organization for local inventor organizations around the country, and has actively worked to build and strengthen the infrastructure of the independent inventor community. They have created a comprehensive database for inventor groups and organizations around the country. Visit www.uiausa.org/Resources/Inventor-Groups.htm

U.S. Patent and Trademark Depository Libraries (PTDL)
These libraries fall under the management of the United States Patent and Trademark Office, and are helpful for conducting patent and trademark searches. Some allow teleconferences directly with the USPTO. The PTDLs have librarians on staff who will help you conduct patent searches, and will answer questions related to your patent filing. For more information, call 800-PTO-9199, or see the full list of PTDLs online at www.uspto.gov/web/offices/ac/ido/ptdl/index.html

About the Authors

Bob DeMatteis is the inventor-marketer of twenty U.S. patents that have been licensed and successfully commercialized. He is the founder of the From Patent to ProfitÆ and the Patents in Commerce training series, two programs designed to assist fellow inventors. He is also a Certified Seminar LeaderÆ and an advisor to several small company development organizations. Mr. DeMatteis has written several books in the area of invention, including the bestseller *From Patent to Profit*.

Andy Gibbs is the founder and CEO of PatentCafe. He was twice appointed by the U.S. Secretary of Commerce to the U.S. Patent and Trademark Office Public Patent Advisory Committee. His past twenty-five years of business experience ranges from being a VP of a Fortune 500 company to founding seven companies. He is also a product developer and holds nearly a dozen patents. Mr. Gibbs is the author of over 100 articles on patents and intellectual property, and the coauthor of *Essentials of Patents*.

Michael Neustel is a U.S. Registered Patent Attorney with a bachelor of science in Electrical Engineering. He is licensed to practice in front of the U.S. Patent and Trademark Office (USPTO), and is the founder of the National Inventor Fraud Center, Inc. He regularly presents intellectual property seminars for various inventor organizations throughout the United States, and is the creator of popular software products such as PatentWizard and PatentHunter.

Index

FROM PATENT TO PROFIT
Secrets & Strategies for the Successful Inventor
Bob DeMatteis

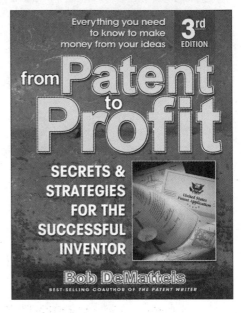

So you've come up with an invention. Now what? Having a novel idea and turning that idea into cash is not as simple as it sounds. The process of inventing a product, protecting that invention, and bringing it to the marketplace can be costly in both time and money. In today's highly complex and competitive world of business, not knowing what to expect and what to do is a certain guarantee of frustration and failure.

To help innovators and small businesses learn to navigate around the many pitfalls of inventing, Bob DeMatteis has written From Patent to Profit, an up-to-date guide to all of the important steps involved in taking a product from drawing board to market. As someone with hands-on experience in every phase of the inventing process—creation, patenting, licensing, manufacturing, and marketing—DeMatteis has made his share of mistakes, and has also learned from them. He has taken his accumulated knowledge and turned it into a well-organized handbook with proven methodologies that have resulted in the birth of many successful products. To further guide readers, he shares the insights and advice of successful inventors and marketers, such as Walt Disney, Jerome Lemelson, Thomas Edison, Yujiro Yamamoto, Edwin Land, and Bill Gates.

All of the information presented in this book is reliable, comprehensive, and easy to understand. The guidance it offers will allow anyone with a creative streak or any small business to sail around potential problems and set a course towards a successful launch. Whether you are a professional inventor, an engineer, or a part-time dabbler, From Patent to Profit can help make your dreams a reality.

ABOUT THE AUTHOR
Bob DeMatteis is the inventor-marketer of more than twenty U.S. patents that have been licensed and successfully commercialized. He is the founder of the From Patent to Profit ® and the Patents in Commerce™ training series, two programs designed to help and assist fellow inventors. He is also an adviser to several small companies and government agencies. Mr. DeMatteis has cowritten several best-selling books in the area of invention, including *The Patent Writer*.

$29.95 US / $44.95 CAN • 432 pages • 8.5 x 11-inch quality paperback • Technology/Reference • ISBN 0-7570-0140-8